Carbon Management: Implications for R&D in the Chemical Sciences and Technology

A WORKSHOP REPORT
TO THE
CHEMICAL SCIENCES ROUNDTABLE

Chemical Sciences Roundtable

Board on Chemical Sciences and Technology

Division of Earth and Life Studies

National Research Council

NATIONAL ACADEMY PRESS
Washington, D.C.

NOTICE: The project that is the subject of this report was approved by the Governing Board of the National Research Council, whose members are drawn from the councils of the National Academy of Sciences, the National Academy of Engineering, and the Institute of Medicine.

Support for this project was provided by the National Science Foundation under Grant No. CHE-9630106, the National Institutes of Health under Contract No. N01-OD-4-2139, and the U.S. Department of Energy under Grant No. DE-FG02-95ER14556. Any opinions, findings, conclusions, or recommendations expressed in this material are those of the authors and do not necessarily reflect the views of the National Science Foundation, the National Institutes of Health, or the U.S. Department of Energy.

International Standard Book Number 0-309-07573-4

Additional copies of this report are available from:

National Academy Press
2101 Constitution Avenue, NW
Box 285
Washington, DC 20055
800-624-6242
202-334-3313 (in the Washington metropolitan area)
http://www.nap.edu

Board on Chemical Sciences and Technology
2101 Constitution Avenue, NW
NAS 273
Washington, DC 20418
202-334-2156

Copyright 2001 by the National Academy of Sciences. All rights reserved.

Printed in the United States of America

THE NATIONAL ACADEMIES

National Academy of Sciences
National Academy of Engineering
Institute of Medicine
National Research Council

The **National Academy of Sciences** is a private, nonprofit, self-perpetuating society of distinguished scholars engaged in scientific and engineering research, dedicated to the furtherance of science and technology and to their use for the general welfare. Upon the authority of the charter granted to it by the Congress in 1863, the Academy has a mandate that requires it to advise the federal government on scientific and technical matters. Dr. Bruce M. Alberts is president of the National Academy of Sciences.

The **National Academy of Engineering** was established in 1964, under the charter of the National Academy of Sciences, as a parallel organization of outstanding engineers. It is autonomous in its administration and in the selection of its members, sharing with the National Academy of Sciences the responsibility for advising the federal government. The National Academy of Engineering also sponsors engineering programs aimed at meeting national needs, encourages education and research, and recognizes the superior achievements of engineers. Dr. William A. Wulf is president of the National Academy of Engineering.

The **Institute of Medicine** was established in 1970 by the National Academy of Sciences to secure the services of eminent members of appropriate professions in the examination of policy matters pertaining to the health of the public. The Institute acts under the responsibility given to the National Academy of Sciences by its congressional charter to be an adviser to the federal government and, upon its own initiative, to identify issues of medical care, research, and education. Dr. Kenneth I. Shine is president of the Institute of Medicine.

The **National Research Council** was organized by the National Academy of Sciences in 1916 to associate the broad community of science and technology with the Academy's purposes of furthering knowledge and advising the federal government. Functioning in accordance with general policies determined by the Academy, the Council has become the principal operating agency of both the National Academy of Sciences and the National Academy of Engineering in providing services to the government, the public, and the scientific and engineering communities. The Council is administered jointly by both Academies and the Institute of Medicine. Dr. Bruce M. Alberts and Dr. William A. Wulf are chairman and vice chairman, respectively, of the National Research Council.

CHEMICAL SCIENCES ROUNDTABLE

RICHARD C. ALKIRE, University of Illinois at Urbana-Champaign, *Chair*
MARION C. THURNAUER, Argonne National Laboratory, *Vice Chair*
ALEXIS T. BELL, University of California, Berkeley
DARYLE H. BUSCH, University of Kansas
MARCETTA Y. DARENSBOURG, Texas A&M University
MICHAEL P. DOYLE, Research Corporation
BRUCE A. FINLAYSON, University of Washington
MICHAEL J. GOLDBLATT, Defense Advanced Research Projects Agency
RICHARD M. GROSS, Dow Chemical Company
ESIN GULARI, National Science Foundation
L. LOUIS HEGEDUS, Atofina Chemicals, Inc.
ANDREW KALDOR, Exxon Mobil
FLINT LEWIS, American Chemical Society
MARY L. MANDICH, Bell Laboratories
ROBERT S. MARIANELLI, Office of Science and Technology Policy
TOBIN J. MARKS, Northwestern University
JOE J. MAYHEW, Chemical Manufacturers Association
WILLIAM S. MILLMAN, U.S. Department of Energy
NORINE E. NOONAN, U.S. Environmental Protection Agency
JANET G. OSTERYOUNG, National Science Foundation
NANCY L. PARENTEAU, Organogenesis, Inc.
MICHAEL E. ROGERS, National Institute of General Medical Sciences
HRATCH G. SEMERJIAN, National Institute of Standards and Technology
PETER J. STANG, University of Utah
D. AMY TRAINOR, Zeneca Pharmaceuticals
JEANETTE M. VAN EMON, U.S. Environmental Protection Agency National Exposure Research Laboratory
ISIAH M. WARNER, Louisiana State University

Staff

RUTH MCDIARMID, Senior Program Officer
SYBIL A. PAIGE, Administrative Associate
DOUGLAS J. RABER, Director, Board on Chemical Sciences and Technology
SCOTT C. JENKINS, National Research Council Intern

BOARD ON CHEMICAL SCIENCES AND TECHNOLOGY

KENNETH N. RAYMOND, *Co-Chair*, University of California
JOHN L. ANDERSON, *Co-Chair*, Carnegie Mellon University
JOSEPH M. DESIMONE, University of North Carolina and North Carolina State University
CATHERINE C. FENSELAU, University of Maryland
ALICE P. GAST, Stanford University
RICHARD M. GROSS, Dow Chemical Company
NANCY B. JACKSON, Sandia National Laboratory
GEORGE E. KELLER II, Union Carbide Company (retired)
SANGTAE KIM, Eli Lilly and Company
WILLIAM KLEMPERER, Harvard University
THOMAS J. MEYER, Los Alamos National Laboratory
PAUL J. REIDER, Merck Research Laboratories
LYNN F. SCHNEEMEYER, Bell Laboratories
MARTIN B. SHERWIN, ChemVen Group, Inc.
JEFFREY J. SIIROLA, Eastman Kodak Company
CHRISTINE S. SLOANE, General Motors Research Laboratories
ARNOLD F. STANCELL, Georgia Institute of Technology
PETER J. STANG, University of Utah
JOHN C. TULLY, Yale University
CHI-HUEY WONG, Scripps Research Institute
STEVEN W. YATES, University of Kentucky

Staff

DOUGLAS J. RABER, Director
RUTH MCDIARMID, Program Officer
CHRISTOPHER K. MURPHY, Program Officer
SYBIL A. PAIGE, Administrative Associate

Preface

The Chemical Sciences Roundtable (CSR) was established in 1997 by the National Research Council (NRC). It provides a science-oriented, apolitical forum for leaders in the chemical sciences to discuss chemically related issues affecting government, industry, and universities. Organized by the NRC's Board on Chemical Sciences and Technology, the CSR aims to strengthen the chemical sciences by fostering communication among the people and organizations—spanning industry, government, universities, and professional associations—involved with the chemical enterprise. The CSR does this primarily by organizing workshops that address issues in chemical science and technology that require national attention.

The topic "Carbon Management: Implications for R&D in the Chemical Sciences" was selected by the Chemical Sciences Roundtable in response to concern that the chemical sciences community should be prepared to respond in the event that a policy decision might be implemented in the area of carbon management. The workshop, entitled *Carbon Management: Implications for R&D in the Chemical Sciences,* brought together leaders in chemistry and chemical engineering from government, academia, and industry to gather information and explore possible roles that the chemical sciences R&D community might play in identifying and addressing underlying chemical questions that might arise if government action were taken to regulate carbon dioxide output or fossil fuel consumption. The workshop focused not on the debate over whether we have seen anthropogenically driven climate change or what the climate change effects might be, but on how the chemical community could prepare for and react to a possible national policy of carbon management.

The chapters in this report are the authors' own versions of their presentations, and the discussion comments were taken from a transcript of the workshop. In accord with the policies of the CSR, the workshop did not attempt to establish any conclusions or recommendations about needs and future directions, focusing instead on issues identified by the speakers. The views and opinions of authors expressed herein do not necessarily represent the views of the NRC or any of its constituent units.

Alexis T. Bell and Tobin J. Marks
Workshop Organizers

Acknowledgment of Reviewers

This report has been reviewed in draft form by individuals chosen for their diverse perspectives and technical expertise, in accordance with procedures approved by the (NRC's) Report Review Committee. The purpose of this independent review is to provide candid and critical comments that will assist the institution in making its published report as sound as possible and to ensure that the report meets institutional standards for objectivity, evidence, and responsiveness to the study charge. The review comments and draft manuscript remain confidential to protect the integrity of the deliberative process. We wish to thank the following individuals for their review of this report:

David C. Bonner, Rohm and Haas Company
Glenn A. Crosby, Washington State University
Joseph M. DeSimone, University of North Carolina and North Carolina State University
Gregg Marland, Oak Ridge National Laboratory

Although the reviewers listed above have provided many constructive comments and suggestions, they did not see the final draft of the report before its release. The review of this report was overseen by Edward M. Arnett, Duke University. Appointed by the National Research Council, he was responsible for making certain that an independent examination of this report was carried out in accordance with institutional procedures and that all review comments were carefully considered. Responsibility for the final content of this report rests entirely with the organizers and the institution.

Contents

Summary		1
1	Carbon Management: The Challenge *James A. Edmonds, J. F. Clarke, and J. J. Dooley* (Pacific Northwest National Laboratory)	7
2	Carbon Dioxide Mitigation: A Challenge for the Twenty-First Century *David C. Thomas* (BP Amoco Corporation)	33
3	An Industry Perspective on Carbon Management *Brian P. Flannery* (ExxonMobil Corporation)	44
4	Opportunities for Carbon Control in the Electric Power Industry *John C. Stringer* (Electric Power Research Institute) Session 1 Panel Discussion, 73	60
5	Carbon Dioxide as a Feedstock *Carol Creutz and Etsuko Fujita* (Brookhaven National Laboratory)	83
6	Advanced Engine and Fuel Systems Development for Minimizing Carbon Dioxide Generation *James A. Spearot* (General Motors Corporation)	93
7	Renewable Energy: Generation, Storage, and Utilization *John Turner* (National Renewable Energy Laboratory)	111

8   Industrial Carbon Management: An Overview   127
      David W. Keith (Carnegie Mellon University)
    Session 2 Panel Discussion, 141

9   Managing Carbon Losses for Selective Oxidation Catalysis   147
      Leo E. Manzer (DuPont Central Research and Development)

10 Increasing Efficiencies for Hydrocarbon Activation   159
      Harold H. Kung (Northwestern University)

11 Commodity Polymers from Renewable Resources: Polyactic Acid   166
      Patrick R. Gruber (Cargill Dow LLC)

12 Chemicals from Plants   185
      John W. Frost, K. M. Draths, David R. Knop, Mason K. Harrup, Jessica L. Barker,
      and Wei Niu (Michigan State University)
    Session 3 Panel Discussion, 197

Appendixes

A   Workshop Participants   211
B   Biographical Sketches of Workshop Speakers   214
C   Origin of and Information on the Chemical Sciences Roundtable   220
D   Acronyms and Definitions   222

Summary

Alexis T. Bell
University of California at Berkeley
and
Tobin J. Marks
Northwestern University

Considerable international concerns exist about global climate change and its relationship to the growing use of fossil fuels. Carbon dioxide is released by chemical reactions that are employed to extract energy from fuels, and any regulatory policy limiting the amount of CO_2 that could be released from sequestered sources or from energy-generating reactions will require substantial involvement of the chemical sciences and technology R&D community.

Much of the public debate has been focused on the question of whether global climate change is occurring and, if so, whether it is anthropogenic, but these questions were outside the scope of the workshop, which instead focused on the question of how to respond to a possible national policy of carbon management. Previous discussion of the latter topic has focused on technological, economic, and ecological aspects and on earth science challenges, but the fundamental science has received little attention. The workshop was designed to gather information that could inform the Chemical Sciences Roundtable (see Appendix C) in its discussions of possible roles that the chemical sciences community might play in identifying and addressing underlying chemical questions.

OVERVIEW: ECONOMICS AND OTHER DRIVERS FOR CARBON MANAGEMENT

The first session was devoted to setting the context of the workshop—the broad view of the problem, including its magnitude; the motivations for a carbon management policy; the interplay between public, private, and government sectors in the areas of policy; and the strategic issues and options associated with energy production and use, as well as CO_2 separation and sequestration.

James Edmonds, from Pacific Northwest National Laboratory, presented the motivations for carbon management (see Chapter 1). He articulated the theme that carbon management may prove to be one of the greatest challenges of the twenty-first century since, driven by climate change issues, the global energy production and utilization system will have to undergo radical transformation during this period. He suggested that the need to stabilize the atmospheric CO_2 concentration implies that net anthropogenic carbon dioxide emissions—the contributions to the ocean-atmosphere in excess of the CO_2 uptake by other parts of the carbon cycle—*must decline to zero*. This technological premise, in conjunction with the need to carry it out in a cost-effective manner, would have several important near-term implications for the character of efficient policy development. Using the premises that there are sufficient economic fossil fuels to last for the next century and that continued economic development will ensure a continued growth in energy demand, he described existing patterns for world use of low-cost fossil fuel and identified the need for a portfolio of technologies that will change those patterns. He argued that investments to develop those technologies and their associated infrastructure will require funding a full spectrum of R&D, coordinated among many nations and with the participation of both public and private sectors. However, he showed that global energy R&D has declined over the last 15 years. In his view, public policy will play an important role in signaling the need for new technologies and in facilitating their development and deployment.

David Thomas of BP Amoco discussed various options for CO_2 mitigation and presented the approach that BP has taken (see Chapter 2). He presented examples of methods for reducing energy consumption from various manufacturing processes practiced by BP and described possible separation technologies for the CO_2 that is emitted. Separation of CO_2 could occur early, in capture from natural gas, or much later, from combustion processes. He also described options for storing CO_2 safely, particularly in geological formations. Thomas reiterated Edmonds's point that mitigation techniques must be cost-effective if they are to be widely adopted.

Brian Flannery of ExxonMobil (see Chapter 3) pointed out that concerns about anthropogenic emissions of CO_2 and their possible effect on climate change have led to policy proposals that would dramatically restrict future emissions. He revisited some of the issues discussed by Edmonds and provided a historical perspective on energy use, energy decarbonization and energy efficiency. His talk emphasized the magnitude of the problem and time scale for penetration of new technologies. He presented several scenarios that could lead to CO_2 stabilization at various levels over the next century, noting that the associated social, environmental, and economic costs would be sensitive to the availability and performance of new technologies. He emphasized that effective implementation of any new energy technology would require extensive infrastructure development, and that any policy requiring stabilization of atmospheric CO_2 concentrations would depend on development and widespread global implementation of technologies that are not commercially available today. He identified a twofold chemical R&D focus that would enhance our ability to assess the extent and consequence of climate change and would contribute to the development of appropriate advanced technologies.

Opportunities for carbon emissions control in the electric power industry were addressed by John Stringer of the Electric Power Research Institute (see Chapter 4). He emphasized that the problem is very large, that the dominant worldwide generation fleet will increasingly consist of fossil fuel-fired thermal stations, and that no clearly superior methods for carbon management currently exist. On that basis, he concluded that research on multiple candidates—including combustion systems, nuclear energy, and renewable energy—will be critically important. He highlighted the importance of having a long-range roadmap for planning the supply of electricity for the next 50 years, since most of the world's current generating capacity would be replaced during that period. He described options for reducing CO_2 emissions associated with power generation and discussed the issues associated with the principal fuels: petroleum, natural gas, and coal. He summarized opportunities for CO_2 capture, particularly for coal-fired plants and identified the issues associated with alternative sequestration strategies. Finally, he reiterated Edmonds's point that a commitment to emissions mitigation should not be made too soon, lest the approach prove unsuitable or unachievable at a reasonable cost.

In the panel discussion following the first session, an effort was made to focus on identifying a research agenda in chemical sciences and engineering that would be aimed at reduction of CO_2 emissions. Nevertheless, the majority of the discussion concerned the economics of carbon mitigation, and participants reported that simple steps, such as fuel switching, are occurring. Several participants expressed the need for creative and innovative research to lay the foundations for the next generation of technology—as argued by Brian Flannery, current approaches appear unlikely to succeed in the near term.

SCIENCES R&D ISSUES IN MANAGING PRODUCED CARBON DIOXIDE

The presentations in the second session of the workshop were centered on the efficient use of carbon resources that could lead to a reduced contribution to the CO_2 pool. The presentations covered a number of perspectives, ranging from laboratory research to industrial policy. Strategies suggested to accomplish the goals ranged from a system of renewable fuels that could avoid CO_2 emissions to a system that would rely on fossil fuels with separation and sequestration of the CO_2.

Carol Creutz of Brookhaven National Laboratory reviewed the use of carbon dioxide as a starting material for organic synthesis, for potential industrial chemical applications, and as a feedstock for fuel production (see Chapter 5). She put this in perspective by comparing the estimated net anthropogenic increase of 13,000 million tons of CO_2 added to the atmosphere annually with the annual total of 110 million tons transformed into chemicals—mainly urea, salicylic acid, cyclic carbonates, and polycarbonates. She suggested that increased use of CO_2 as a starting material would be desirable, potentially producing a positive—although small—impact on global CO_2 levels. Use of supercritical CO_2, a hydrophobic solvent that can replace organic solvents in a number of applications, could consume additional amounts. Reactive use of CO_2 is limited by the fact that it is very stable, and energy must be supplied to drive most transformations. Dr. Creutz suggested that renewable energy sources be considered in driving CO_2 utilization, such as direct hydrogenation to CH_3OH or CH_4 via a variety of routes, including photochemistry. She concluded her presentation by identifying several areas of ongoing and future research directions that could lead to CO_2 utilization in new polymers.

James A. Spearot of General Motors discussed advanced engine and fuel systems for minimizing CO_2 generation (see Chapter 6). The goal of the automotive industry is to respond to the global demand for the freedom provided by modern transportation technology and thereby achieve sustainable "auto-mobility." He suggested that for society to continue to enjoy the benefits of personal mobility, we will need long-term energy forms that are renewable and vehicle technologies that have zero impact on the ambient environment. He discussed reduction of carbon emissions by improving the efficiency of vehicles and propulsion systems—through the auto industry-government program known as the Partnership for a New Generation of Vehicles (PNGV). Various improvements in fuel economy might be obtained through use of lightweight bodies, advanced combustion technologies, and advanced transmissions. In addition to combustion engines, fuel cell-powered vehicles operating with either gasoline or hydrogen could provide significant efficiency improvements and CO_2 emission reductions, but the penetration of hydrogen-fueled vehicles would require the development of a hydrogen infrastructure. Spearot reported that significant progress has been made, but he cautioned that future emission standards, particularly for NO_x and particulates, might limit utilization of some of the near-term advanced technologies. Advanced propulsion systems also will require advanced fuel compositions, and the use of hydrogen—the ultimate fuel—will require significant advances in on-board hydrogen storage as well as the development of a fuel delivery infrastructure. Spearot also suggested that biomass-based ethanol could represent a CO_2 conservation option, but this was challenged by Flannery, who earlier had presented a different perspective (see Chapter 3).

John Turner of the National Renewable Energy Laboratory discussed renewable energy storage, generation, and utilization (see Chapter 7). He proposed that advanced renewable energy systems may provide the basis for a sustainable energy supply without net anthropogenic emission of CO_2. He argued that large-scale implementation of renewable energy technologies could eliminate the need for CO_2 sequestration by reducing the use of—and ultimately eliminating the need for—fossil-based energy production. The renewable energy systems he discussed include photovoltaics, solar thermal (electric and thermal), wind, biomass (plants and trees), hydroelectric, ocean, and geothermal. He described the impressive growth of wind power, particularly on wind farms. He emphasized the need for energy storage technologies that would overcome the intermittent nature of several of the renewable energy sources, as well as the need for basic research in all aspects of renewable energy generation. One option he emphasized for energy storage was electrolysis for renewable hydrogen generation.

David W. Keith of Carnegie Mellon University (Chapter 8) spoke about industrial carbon management to permit the continued use of fossil fuels for energy. Industrial carbon management links processes for capturing the carbon content of fossil fuels while generating carbon-free energy products such as electricity and hydrogen and sequestering the resulting carbon dioxide. The energy content of the fossil fuels would first be separated from their carbon content in one of three separation schemes: post-combustion capture (combustion in air followed by removal of CO_2 from the combustion products), oxyfuel (separation of oxygen from air followed by combustion in pure oxygen and CO_2 capture), and pre-combustion decarbonization (with a first step of reforming the fuel to produce hydrogen and CO_2 followed by capture of the CO_2). The third alternative has the advantage that a power plant could sell zero-CO_2-emission hydrogen for a hydrogen infrastructure. In all cases, CO_2 sequestration is the substantive challenge. For such technologies Keith estimated the cost of electricity to be about 2-3 cents per kilowatt-hour more than from current technologies, roughly comparable with the cost of electricity via wind, biomass, or nuclear power. He argued that the advantage of industrial carbon management is its fit with the existing infrastructure for power generation and distribution, and he suggested that a carbon tax might provide the economic driver to accelerate this approach.

The discussion following this session brought up the issue of biological sequestration of carbon. David Thomas questioned whether regional politics might influence national decisions on sequestration, and David Keith suggested that biological sequestration could have a significant short-term impact. There was extensive discussion among the panel and participants on the costs and challenges for hydrogen storage and renewable energy sources.

EFFICIENT UTILIZATION OF CARBON RESOURCES

The third session of the workshop was devoted to the efficient utilization of carbon resources. Two perspectives were presented: the first was devoted to efficient utilization of hydrocarbon resources in "traditional" chemistry, and the second focused on using plants as the feedstock for making chemicals and polymers.

Leo Manzer of DuPont (Chapter 9) described how selective industrial catalytic oxidation could be used to reduce carbon losses from processes for making olefins and oxygenated products. He began by noting that catalytic oxidations usually exhibit selectivities for desired products of less than 90% and account for much of the CO_2 released from chemical processes. He illustrated how reductions in CO_2 emissions could be achieved by several means, including a two-step oxidation process in which a hydrocarbon is oxidized anaerobically by the catalyst and the partially reduced catalyst is then reoxidized in a separate step. Such processes have been found to yield higher product selectivities than those obtained from aerobic oxidation. Higher product selectivities can also be achieved by using alternative oxidants such as hydrogen peroxide (H_2O_2) and nitrous oxide (N_2O), but the cost of these oxidants would have to be reduced significantly before they could become commercially attractive compared to oxygen. He described the potential use of hydrogen-oxygen mixtures for selective oxidation but noted that serious safety issues would have to be addressed. Manzer also showed how creative new catalytic technologies could be used to achieve higher product yields and reduced CO_2 emissions. He also suggested that the use of CO_2 as a mild oxidant is an interesting new development to be pursued. However, he argued that it is not economical to replace existing chemical plants, so financial incentives would be needed to commercialize improved technologies.

The theme of efficient utilization of hydrocarbon resources was continued by Harold Kung from Northwestern University (Chapter 10), who noted that oxidation catalysts exhibiting higher selectivities are needed for processes such as the oxidative dehydrogenation of alkanes to olefins and the selective oxidation of alkanes to oxygenated products. Kung observed that the hydrogen-to-carbon ratio of most petrochemicals is higher than that in crude oil, so hydrogen must be added to make these products. Since hydrogen is produced by the steam reforming of methane, a process that generates CO_2, reducing the consumption of hydrogen would lead to a reduction in CO_2 emissions. He illustrated several alternatives by which oxygenated products might be produced to minimize the loss of hydrogen as water, and he urged the development of novel processes that minimize the consumption of energy. For example, if acetic acid could be produced by direct heterogeneous carbonylation of methanol, then the energy required to separate acetic acid from the water solvent, a considerable component of the energy consumption, could be avoided. Opportunity also exists to develop new strategies for reforming liquid fuels to produce hydrogen for fuel cells.

The production of polylactic acid (PLA) from cornstarch was addressed by Patrick Gruber of Cargill Dow (see Chapter 11). PLA was chosen as a target for process development because it exhibits adequate performance as a commodity polymer and would command a reasonable market price. Since CO_2 is fixed in crops to make starch, the starting material for PLA comes from a renewable source. Gruber addressed the market opportunities and potential for PLA and the relationship between the production of PLA and the net consumption or production of CO_2. Currently there is market demand for PLA in three areas: fibers, packaging, and chemical products. Long-range opportunities also exist for converting lactic acid, the monomer for PLA, into a variety of commodity and specialty chemicals, as well as polymers other than PLA. Gruber projected that in its first year of operation, the cradle-to-pellet emission of CO_2 for the production of PLA would be comparable to that for the production of polypropylene, but with further process development, PLA production would become a net consumer of CO_2. His overall conclusion was that products made from renewable resources, such as PLA, offer advantages over the petrochemical-based products that they would replace, and he argued that the development of successful renewable resource-based products will require the skillful combination of expertise in fermentation and biotechnology with that in more conventional chemistry and polymer science.

John Frost of Michigan State University described the production of chemicals from plants (see Chapter 12). He began with the observation that virtually all commercial chemicals are synthesized from petroleum feedstocks; very few are isolated as natural products from plants or produced microbially from plant feedstocks. He argued that there is enormous opportunity for the production of chemicals from renewable feedstocks and, hence, a net consumption of CO_2, since plants can be thought of as immobilized forms of CO_2. His strategy is based on the development of new syntheses and synthetic methodologies compatible with the use of water as the reaction solvent. Also critical would be effective interfacing of microbial catalysis with chemical catalysis and the discovery and use of genes encoding biosynthetic enzymes. Frost illustrated the opportunities for his approach with a number of examples. Shikimic acid is a key intermediate in the synthesis of Tamiflu, an effective anti-influenza drug, but is available in only limited quantities. He showed that shikimic acid could be synthesized from readily available starting materials via a biosynthetic route. Examples were presented also of how one might synthesize hydroquinone and adipic acid from glucose. Frost ended his presentation with a call for chemists to view construction of microbial catalysts as an activity every bit as central to chemical synthesis as the development of inorganic and organometallic catalysts. Both Frost and Gruber emphasized the need for broadly trained scientists who understand chemistry and can move across disciplinary boundaries.

The panel discussion following the third session began with the question of how the issues presented by the speakers could be translated into a research agenda for the chemical sciences. In response to questions about the use of computer modeling, the speakers indicated that this was an important component of industrial research, but it is just one component. Several participants asked if the increasing national emphasis on biologically related research might undermine progress in the physical sciences. Manzer and Gruber indicated that industry needs scientists who can work across the boundaries of the disciplines, and others suggested that future success in carbon management would require broad efforts at understanding the chemistry of CO_2 in both biological and geological contexts.

The contributions in this report from the workshop speakers indicate that a program of carbon management would pose enormous challenges. Several speakers described ways that R&D could reduce the amount of CO_2 that is generated by chemical industry. While it was noted that this is only a small fraction of the total amount of anthropogenic CO_2 (see the discussion following Chapter 2), reductions could be economically important to the chemical industry if a carbon management policy were to be established. Several speakers pointed to ways that R&D in the chemical sciences and engineering might lead to reduction of emissions by the power and transportation sectors, which are responsible for the preponderance of CO_2 generated by human activity.

1

Carbon Management: The Challenge

James A. Edmonds, J. F. Clarke, and J. J. Dooley
Pacific Northwest National Laboratory[1]

The Framework Convention on Climate Change (FCCC) was negotiated in Rio de Janeiro in 1992 and entered into force on March 21, 1994. It has been ratified by 186 parties, including the United States.[2] The ultimate objective of the FCCC is set forth in Article 2:

> *The ultimate objective of this Convention and any related legal instruments that the Conference of the Parties may adopt is to achieve, in accordance with the relevant provisions of the Convention, stabilization of greenhouse gas concentrations in the atmosphere at a level that would prevent dangerous anthropogenic interference with the climate system. Such a level should be achieved within a timeframe sufficient to allow ecosystems to adapt naturally to climate change, to ensure that food production is not threatened and to enable economic development to proceed in a sustainable manner.*

In addition, Article 3 of the FCCC specifies that, ". . . policies and measures to deal with climate change should be cost-effective so as to ensure global benefits at the lowest possible cost."

The FCCC established many important principles but left most of the critical details to be worked out by later conferences of the parties to the convention. For example, the objective of the convention is the stabilization of the concentration of greenhouse gases, but the level at which they should be stabilized is not specified. There is as yet no scientific basis for preferring one concentration limit to another,

---

[1] *Disclaimer:* This report was prepared as an account of work sponsored by an agency of the U. S. government. Neither the U. S. government or any agency thereof, nor Battelle Memorial Institute, nor any of their employees makes any warranty, express or implied, or assumes any legal liability or responsibility for the accuracy, completeness, or usefulness of any information, apparatus, product, or process disclosed, or represents that its use would not infringe privately owned rights. Reference herein to any specific commercial product, process, or service by trade name, trademark, manufacturer, or otherwise does not necessarily constitute or imply its endorsement, recommendation, or favoring by the U.S. government, any agency thereof, or Battelle Memorial Institute. The views and opinions of authors expressed herein do not necessarily state or reflect those of the U.S. government or any agency thereof. (Pacific Northwest National Laboratory, *operated by* Battelle Memorial Institute *for the* United States Department of Energy *under Contract DE-AC06-76RLO 1830.*)

[2] As such it differs from its more controversial cousins such as the Kyoto Protocol, which was negotiated in 1997 under the FCCC, but has not entered into force.

and there likely will never be.[3] Preindustrial concentrations of carbon dioxide, for example, were in the neighborhood of 275 parts per million volume (ppmv). They had risen to 368 ppmv by 1999. Under a variety of scenarios this concentration rises to anywhere from 500 to more than 700 ppmv over the course of the twenty-first century.[4]

The natural carbon cycle governs the relationship between emissions and concentrations. Anthropogenic emissions originating from net changes in land-use and fossil fuel oxidation initially enter the atmosphere but are eventually partitioned between the atmosphere and the ocean. While the oceans ultimately take up much of the net release, a fraction of any net emission remains in the atmosphere for more than a millennium. As a consequence, the preindustrial level of 275 ppmv concentration of CO_2 is no longer accessible in the present millennium without reversing the net flow from fossil fuel oxidation and land-use change.

Stabilizing the concentration of CO_2 in the atmosphere therefore implies that net emissions ultimately fall to zero from present levels, which are in excess of 6 petagrams of carbon per year (PgC/yr).[5] It is cumulative emissions that matter. Therefore, the particular time path of emissions will depend on the concentration to which atmospheric CO_2 is limited. It also means that to satisfy the cost-effectiveness objective of the FCCC, emissions may rise before finally declining. Emissions trajectories consistent with five alternative concentration limits are shown in Figure 1.1. Key characteristics associated with emissions paths consistent with CO_2 concentrations limits are shown in Table 1.1.

FACTORS SHAPING FUTURE GLOBAL CARBON EMISSIONS

Understanding the key drivers of historic and future carbon emissions is critical to developing policies to control emissions. Yoichi Kaya developed a simple analytical framework to understand the relationship between population, economic activity, energy, and emissions. He observed a simple identity that has significant analytical power:

$$C = (C/E)*(E/Y)*(Y/N)*N.$$

In this equation, C = carbon emissions per year, E = energy consumption per year, Y = GNP per year,[6] and N = population. It implies that the rate of change in carbon emissions is simply the sum of the rates of change of carbon intensity (C/E), energy intensity (E/Y), per capita income (Y/N), and population growth.

By using the above equation, historical trends can be examined and future analysis dissected. Figures 1.2 through 1.5 show population, GNP per capita, energy intensity, and carbon intensity for historical and forecast years. Data for forecast years are taken from Pepper et al. (1992), which docu-

---

[3]There are multiple greenhouse-related gases. These include water vapor, carbon dioxide, carbon monoxide, methane, nitrous oxide, odd nitrogen compounds, the chlorofluorocarbons and their replacements, and aerosol compounds. Carbon dioxide is the most important human-released greenhouse gas from the perspective of potential change in future climate. Its principal source of emissions is fossil fuel use, however, land-use change in general and deforestation in particular also play important roles.

[4]See for example IPCC (1996a) and Nakicenovic et al. (2000), although the latter gives only emissions and cumulative emissions calculations.

[5]1 Pg = 1 billion tonnes.

[6]In this equation GNP refers to "gross world product." Therefore we use GNP to mean the value of all new final goods and services produced in the world in a given year. We continue to use the acronym GNP to avoid confusion with the concept of global warming potential (GWP), which is also found in the literature.

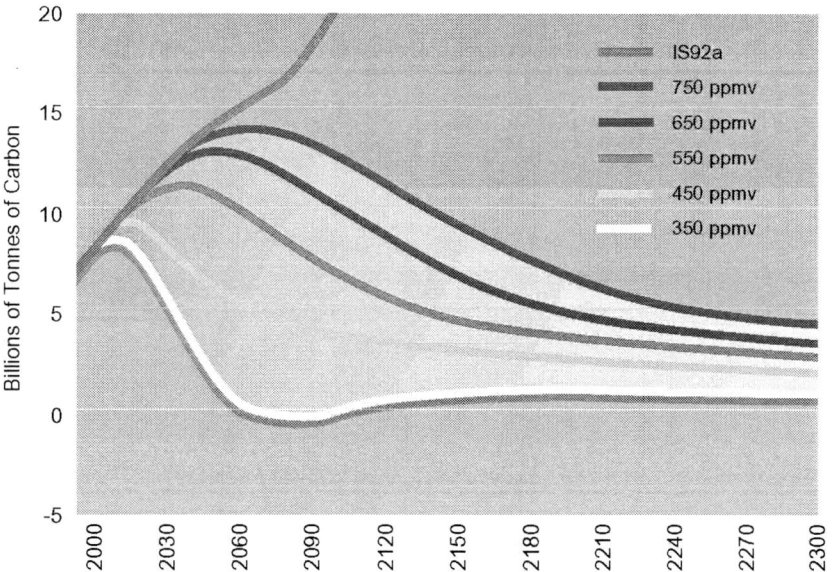

FIGURE 1.1 Net annual anthropogenic carbon emissions paths consistent with alternative CO_2 concentration limits and the assumed population growth from the IS92a scenario of the Intergovernmental Panel on Climate Change. The curve marked IS92a assumes no CO_2 concentration limit. The specified limit occurs in the year 2150.

ments the Intergovernmental Panel on Climate Change (IPCC) IS92 scenario series.[7] Historical declines in carbon intensity and energy intensity have been more than offset by increases in population and GNP per capita. IPCC carbon emissions scenarios shown in Figures 1.2-1.5 span a range of trajectories. Most cases exhibit higher carbon *emissions* at the end of the twenty-first century than in 1990. All cases exhibit higher CO_2 *concentrations* at the end of the twenty-first century than in 1990. The range of

TABLE 1.1 Characteristics of Potential Emissions Trajectories That Limit Cumulative Atmospheric CO_2 Emissions

	Ceiling (ppmv)				
	350	450	550	650	750
Date when emissions are lower in the control case than in the reference case, IPCC IS92a	Today	2007	2013	2018	2023
Maximum global emission (PgC/yr)	6.0	8.0	9.7	11.4	12.5
Year of maximum global emissions	2005	2011	2033	2049	2062
Annual rate of long-term emissions decline (%)	—	1.1	0.8	0.6	0.5
Cumulative emissions, 1990 to 2100 (PgC)	363	714	1,043	1,239	1,348

[7]The IS92 scenarios are a set of assumptions, the more significant of which concern population, economic growth, rate of end-use energy intensity improvement, and elasticity of demand for energy in developing nations.

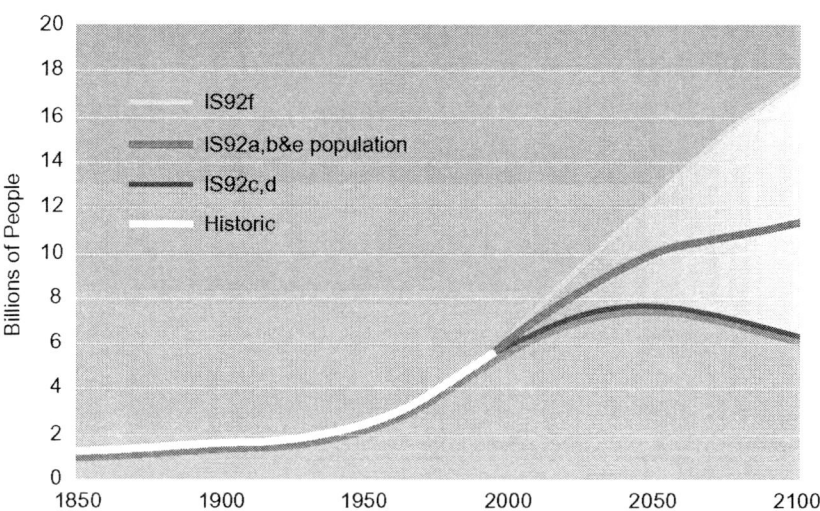

FIGURE 1.2 Global population scenarios.

emissions scenarios developed by Nakicenovic et al. (2000) is similar to that explored by Pepper et al. (1992), in Figure 1.6.

Contrary to popular belief,[8] there is no practical resource limit to human ability to load fossil fuel carbon into the atmosphere. While the resource base of conventional oil and gas is limited, the amount of carbon stored in fossil fuels is not. Table 1.2 describes the distribution of carbon in fossil fuel resources. The amount of carbon stored in the form of conventional oil and gas is only about half the mass of carbon existing in the atmosphere. The amount of carbon stored in the form of coal resources could exceed the amount of carbon in the atmosphere by as much as an order of magnitude. Further, the

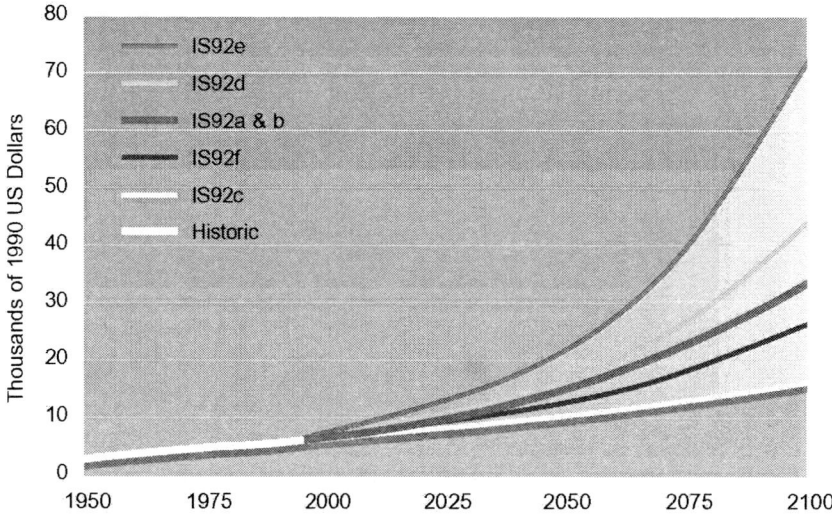

FIGURE 1.3 Global GNP per capita under various scenarios.

---

[8]See for example Kaku (1997, p. 277)—"Within the next thirty years, fossil fuels will become increasingly scarce and prohibitively expensive."

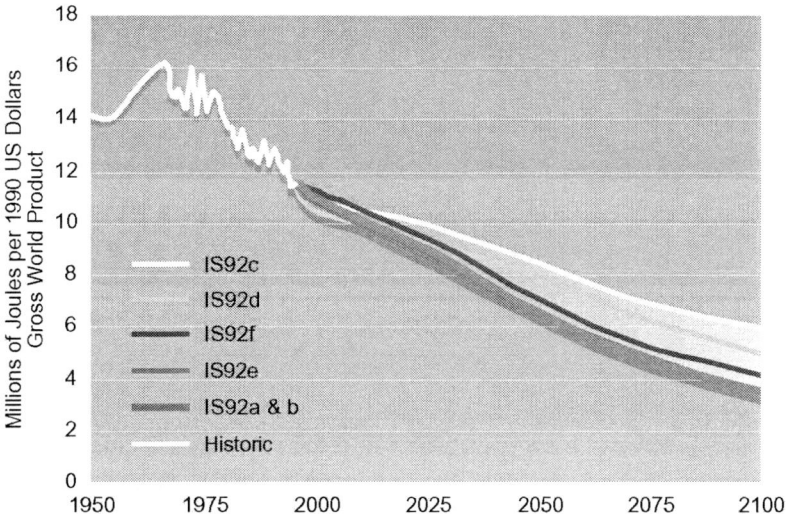

FIGURE 1.4 Energy-GNP ratio under different scenarios.

amount of carbon stored in the form of unconventional liquids and gases exceeds the carbon stored in the form of coal. There is no serious prospect for "running out" of fossil fuels during the course of the twenty-first century. Therefore, the idea that society will soon develop noncarbon energy forms to provide for growing energy demands, because there is no fossil fuel alternative, is unlikely.

TECHNOLOGY IN A REFERENCE FUTURE

Energy technology is assumed to evolve dramatically in reference scenarios, a feature that can go unappreciated. The IS92a scenario, for example, assumes that noncarbon energy forms are sufficiently inexpensive that by the end of the twenty-first century, approximately 75% of all electricity is provided by non-carbon-emitting sources. Similarly, commercial biomass is assumed to grow to the point that it

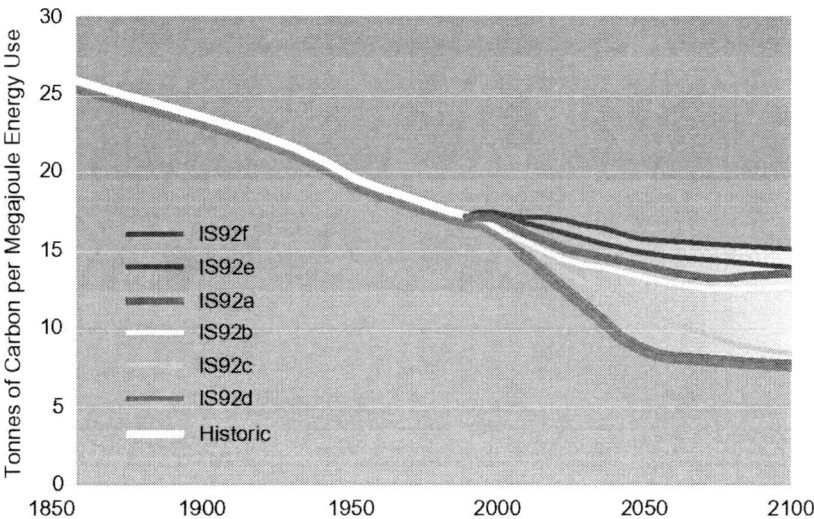

FIGURE 1.5 Global carbon intensity under various scenarios.

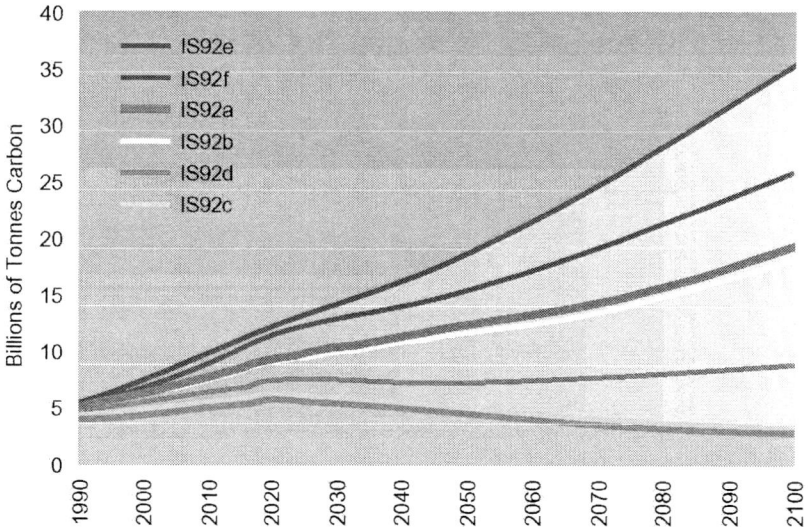

FIGURE 1.6 Range of IPCC IS92 anthropogenic carbon emissions under various scenarios.

provides more energy than did all oil and gas production in the year 1990. This all happens in a "business-as-usual" scenario in which there is no explicit policy to address climate change. At the same time, end-use energy intensity continues to fall to less than half of the level in 1990. To understand the impact of these technological changes on the reference case, we have calculated the emissions under static 1990 technology assumptions and compare these with the reference scenario IS92a. The results of this calculation are shown in Figure 1.7.

The impact of assumed technological change on carbon emissions is huge. Yet the reduction in emissions brought about by these expected developments is insufficient to stabilize CO_2 concentrations. Figure 1.7 shows a path that stabilizes CO_2 concentrations at 550 ppmv. The selection of this concentration is arbitrary. As discussed earlier, other concentration limits could as easily have been selected. The

TABLE 1.2 Carbon Content of Fossil Fuel Energy Resources Potentially Available After 1990

Energy Form	Resource Base (PgC)	Range of Resource Base Estimates (PgC)	Additional Occurrences (PgC)	Resources plus Additional Occurrences (PgC)
Conventional oil[a]	170	156-230	200	156-430
Conventional gas[a]	140	115-240	150	115-390
Unconventional gas[a]	410	—	340	750
Coal[a,b,c]	3,240	—	3,350	3,240-6,590
Tar sands and heavy oils[e,d]	720	600-800	—	600-800
Oil shale[c,e]	40,000	—	—	40,000
Gas hydrates[a]	—	—	12,240	12,240

[a]IPCC (1996b p. 87).
[b]Assumes 50% unrecoverable coal in the resource base.
[c]Range estimates are not available due to the abundance of the resource.
[d]Rogner (1996).
[e]Edmonds and Reilly (1985).

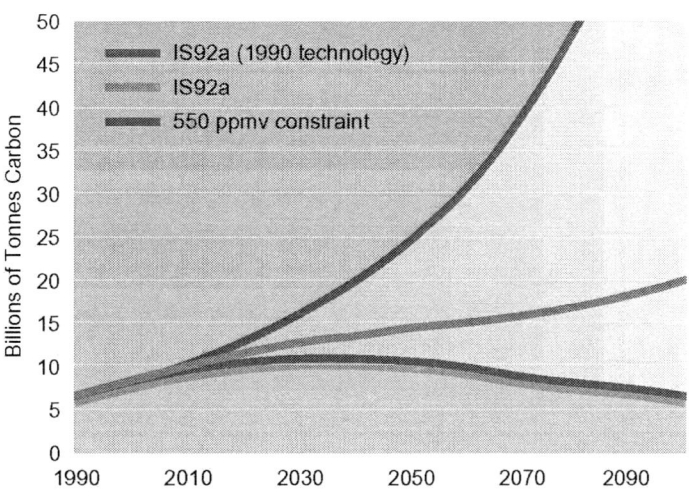

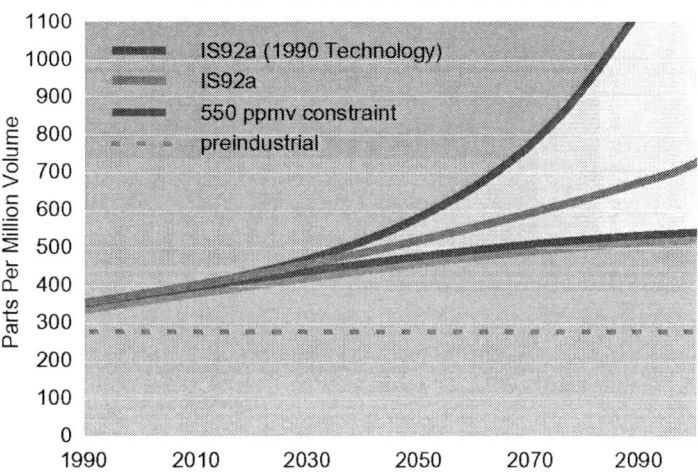

FIGURE 1.7 Comparison of carbon emissions and CO_2 concentrations with and without technological change and compared to a 550-ppmv CO_2 limit.

principal virtue of this particular number is that it happens to be approximately twice preindustrial concentration levels and much work in the climate sciences has employed this value.

The excess of carbon emissions associated with any reference trajectory over those needed to limit the concentration to some prescribed level can be thought of as defining a "gap." Clearly the technology mix must change for that gap to be closed. Either new or improved energy technologies, or both, will be needed to close the gap.

THE ENERGY SYSTEM WITH AND WITHOUT AN EMISSIONS CEILING

To begin to explore the role of technology in filling the gap, we first construct a reference energy system and compare its evolution to an alternative path with a CO_2 concentration limit. We have developed two alternative reference scenarios for global population, economy, energy, and agriculture. The first assumes a transition from the present conventional oil and gas-based world to a future world dominated by coal. This is a standard vision of the future (IPCC, 1995). We call this "coal bridge to the future" (CBF). The second reference case assumes that oil and gas, which are economically unattractive today, become available in the future, thus implying a world that continues to be dominated by oil and gas. We call this "abundant oil and gas" (AOG). This case is also known as "oil and gas future" (OGF).

Both of these energy worlds evolve against a background of continued productivity improvement in energy production, transformation, and end use. These two paths represent the two generic global energy system developments that are interesting from the perspective of climate policy. Either the world can continue to develop using oil and gas as the "backbone" of the global energy system, or it can transition to coal. In both of these cases, other energy forms play important and, in cases such as solar energy, growing roles as the future unfolds. However, the backbone of the world's energy system remains fossil fuels.

The two reference scenarios, CBF and AOG, depicted in Figure 1.8, paint alternative pictures of the overall energy system, although total primary energy and carbon emissions are similar. In CBF, the transition from conventional oil and gas to coal is accompanied by an increase in the price of liquids and gases (due to the higher cost associated with manufacturing coal-based synfuels and syngas) during the first half of the twenty-first century. This leads to lower future energy consumption in end-use applications, but the employment of a greater amount and more carbon-intensive primary energy implies relatively high primary energy demands and carbon emissions.

In the AOG scenario, oil and gas prices remain low because usable resources are never exhausted. The lower energy prices imply a higher level of final energy consumption, but the lower carbon content of the primary energy inputs to the system leaves carbon emissions similar to those of the CBF scenario. We also note that the lower energy prices imply a smaller contribution by renewable energy forms.

Our two reference cases, CBF and AOG, exhibit continued growth in fossil fuel emissions. This growth, in turn, is inconsistent with eventual stabilization of CO_2 concentrations. Because we do not know whether or at what level concentrations will eventually be stabilized, we explore constraints that include 450, 550, 650, and 750 ppmv, as well as reference, (i.e., unconstrained) carbon accumulation.

A third, generic global energy system development path entails an evolution away from fossil fuels toward conservation, renewable, and nuclear energy forms (Shell Oil, 1995). This reference case is uninteresting from the perspective of this chapter because its accomplishment implies the stabilization of greenhouse gas concentrations in the reference case. There are no policy implications. The market takes care of everything.[9]

We compute the cost of achieving each of these alternative objectives under a specific policy regime. This policy regime assumes a coalition of all of the world's nations to mitigate emissions, though there may be compensating transfers of income among nations. At any point in time, we show all economic agents in the model, in all regions. A common value of carbon is to be included in all economic decision making.[10] This strategy minimizes the cost of emissions mitigation at each and every point in time.

---

[9]Although it is possible, we consider a self-stabilizing reference case unlikely. Fossil fuel resources are abundant, and although costs have been highly variable in the short term, they have proved relatively stabile over the long term.

[10]The value of carbon is the premium associated with net carbon emissions that should be employed in all internal planning by energy producers and users to satisfy the associated carbon constraint.

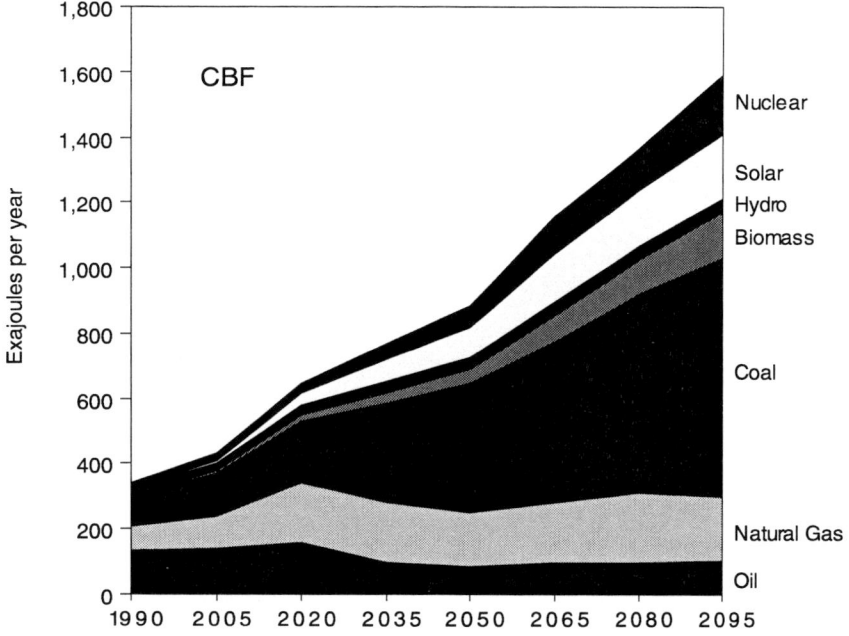

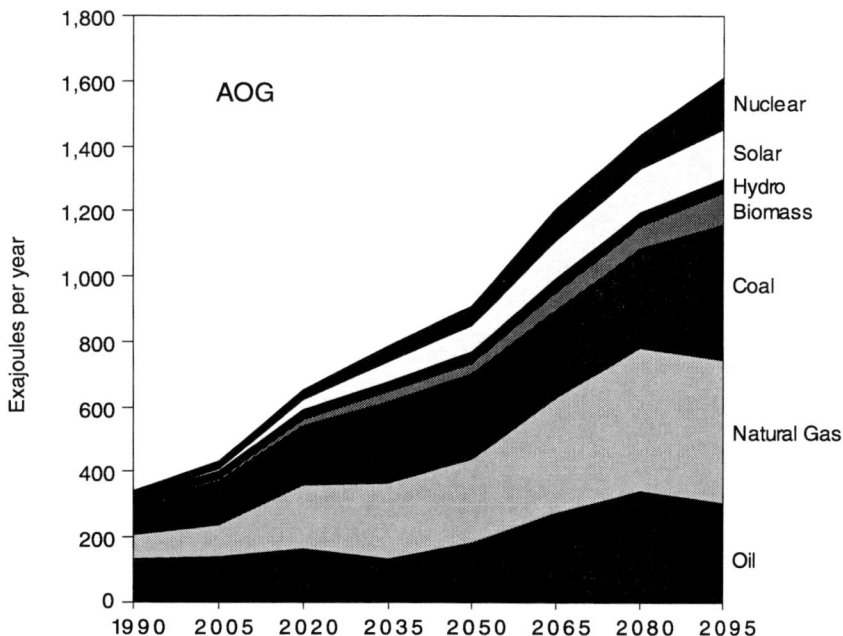

FIGURE 1.8 Reference case energy emissions for the coal bridge to the future (CBF) and abundant oil and gas (AOG) scenarios. AOG is also known as oil and gas future (OGF).

Costs of achieving stable carbon concentrations could be significantly greater if policy interventions undertake expensive emissions mitigation too early in the twenty-first century or if significant portions of the world remain outside the coalition.[11]

Interactions between population, economy, energy, agriculture, land use, greenhouse gas emissions, and atmospheric dispositions are modeled using MiniCAM 98.3. This is an updated version of the MiniCAM model described in Edmonds et al. (1996, 1997). The energy component of the MiniCAM 98.3 has its origins in Edmonds and Reilly (1985). The energy module has been extended and upgraded on numerous occasions. The most recent enhancements are documented in Edmonds et al. (1986) and Kinzey et al. (1998).

TECHNOLOGIES AND THE EMISSIONS GAP

For the gap to be filled with technologies that are not in the reference case, two things must come to pass. First, there must be a credible expectation that the emission of carbon to the atmosphere will be limited. This can be accomplished in a number of ways. Most of the alternatives are economically inefficient. That is, they require more resources to be expended than necessary and thereby foreclose more options than necessary. In the analysis presented here, we assume an economically efficient solution, not because we believe it to be the most likely outcome, but rather because it is unique. There are many ways to be inefficient.

Beyond a commitment to limit carbon concentrations, there must also be a parallel commitment to control cost. Without cost control, the commitment to limit emissions will either flag or be abandoned. The principal beneficiaries of CO_2 concentration limits are not those presently alive, but rather those who will be alive in the future, especially in the second half of the twenty-first century. Also, while altruism is alive and well in the world, like other goods it is in limited supply. If the cost of limiting CO_2 concentrations remains low, both lower concentrations and more current consumption can be had compared with a more costly regime. New and improved technologies that allow emissions of carbon to be controlled are an important source of cost control.

In addition to the advances that are assumed in the reference cases, we introduce several other technologies that are potentially important in a world of CO_2 concentration limitation. These include technologies that allow carbon to be captured and sequestered, capturing the carbon either directly at the source or at sites far removed from the emission and using either terrestrial or geological sinks.

CO_2 Capture from Electric Power Generation

It has long been known that CO_2 can be captured from the exhaust stream of a combustion cycle. However, the capture of CO_2 from the waste stream of a plant requires energy and capital investments. We assume that the efficiency of carbon capture will increase with time (i.e., new and improved technologies and processes will come on-line that reduce the energy penalty associated with powering the capture systems). Herzog et al. (1997) state that the eventual integration of these systems into the overall design of new fossil-fueled power plants—such as integrated gasification combined cycle (IGCC) power plants—holds forth the promise of reducing the cost of CO_2 capture significantly. Further, recent research indicates that targeted basic science programs could lead to advancements that over time would

---

[11]Manne and Richels (1997) have shown that deviation from a cost-effective path can multiply the cost of complying with an atmospheric concentration ceiling such as 550 ppmv. Richels et al. (1996) showed that, if significant regions remain outside an emissions control regime, costs escalate.

improve the performance and reduce the costs of these systems (Dooley and Edmonds, 1997). We assume that the phasing in of these more efficient capture technologies occurs gradually and is completed 50 years after the initiation of carbon capture.

In addition to the energy cost of capture, there are additional capital investment requirements for a CO_2 capture system. Our cost estimates are based largely on the work of Gottlicher and Pruschek (1997) and their comprehensive survey of more than 300 studies of CO_2 removal systems from fossil-fueled power plants. Gottlicher and Pruschek's estimates of the performance of these CO_2 removal systems are based on the "present status of the technology," and therefore we once again adopt the same assumption about costs decreasing over time. Given the wide range of cost reported by Gottlicher and Pruschek (1997), we adopt a midrange cost from their survey (see Table 1.3).

CO_2 Transport Costs

For the foreseeable future, the vast majority of research for disposal applications is likely to be focused on understanding and mitigating environmental concerns that arise from disposal and is unlikely to be directed at reducing the cost of disposal (Freund and Ormerod, 1997). Freund and Ormerod (1997) cite estimates for transport and disposal cost that range from $4.7 to $21 per ton of CO_2 depending upon whether the sequestration is to take place in a nearby depleted oil and gas well or a deep-sea trench that is located some distance from an onshore fossil-fueled power plant. In the absence of research that pairs current and future power plant sites with disposal sites on a global basis, we will assume an intermediate value of $55 per tonne of carbon ($15 per tonne of CO_2) for all transport and disposal costs and hold this cost constant throughout the time period under study.

CO_2 Capture and Disposal Costs for Fuel Transformation Technologies

CO_2 will also have to be captured from fuel conversion facilities such as plants for the conversion of coal to synthetic liquids and gases. Herzog et al. (1997) state that the cost of CO_2 capture from refineries will be comparable to or greater than the cost of capture from fossil-fueled power plants. Therefore, we assume that all conversion facilities and refineries will have performance characteristics similar to those for coal-fired plants. We summarize our assumptions for capture and disposal in Table 1.3.[12] The values

TABLE 1.3 Assumed Cost and Performance of Carbon Capture and Disposal

	Coal	Oil and Gas
Energy penalty for carbon capture[a]	37% declining to 9%	24% declining to 10%
Additional investment costs for capture system[b]	54% declining to 33%	54% declining to 33%
Transport and disposal cost[c]	$55/tonne of C	$55/tonne of C
Efficiency of capture[b]	90%	90%

[a]Herzog et al. 1997.
[b]Gottlicher and Pruschek 1997.
[c]Freund and Ormerod 1997.

---

[12]Note that the figures listed in Table 1.3 for "energy penalty for carbon capture" and "additional investment costs for capture system" are largely consistent with midrange estimates of these parameters published by the International Energy Agency Greenhouse Gas Programme (1996).

in Table 1.3 are not representative of any given capture and sequestration system configuration but rather are meant to be averages for the entire suite of carbon capture and disposal technologies and systems that could be deployed in any number of possible combinations.

Fuel Cell Technology

Fuel cells are a complementary technology to carbon capture and sequestration. They allow hydrogen to be employed as a fuel for transportation, as well as disperse production of heat and power. In the absence of fuel cell technology, the transportation sector depends on batteries and fuels from biomass for emissions mitigation. Fuel cells are assumed to be capable of delivering electricity to both stationary and mobile applications. Our assumptions regarding fuel cell technology are given in Table 1.4.

Soil Carbon

Cole et al. (IPCC, 1996b) estimate that between 40 and 80 Pg of fossil fuel carbon emissions might be offset in existing croplands by applying soil carbon sequestration techniques over the course of the next 50 to 100 years. The estimates for cropland assume the restitution of up to two-thirds of the soil carbon released since the mid-nineteenth century by the conversion of grasslands, wetlands, and forests to agriculture. The experimental record confirms that carbon can actually be returned to soils in such quantities: carbon has been accumulating at rates exceeding 1 tonne per hectare per year in former croplands planted to perennial grasses under the Conservation Reserve Program (CRP) (Gebhart et al. 1994). Soil carbon increases ranging from 1.3 to 2.5 tonne per hectare per year have been estimated in experiments on formerly cultivated land planted to switchgrass, a biomass crop (Brown et al., 1998).

Managed forests, wetlands, and rangelands provide further opportunity for significant carbon storage. For example, when agriculture is converted (or allowed to revert) to forest vegetation in systems with very little management to improve growth, soil carbon may accumulate at rates ranging from near zero to 7 tonnes per hectare per year. In temperate regions, gains range from 0.2 to 0.6 tonne and average about 0.3 tonne per hectare per year (Jenkinson, 1971). Gains in tropical and subtropical forests are greater, ranging from 1.0 to 7.4, with an average of 2.0, tonnes of carbon per hectare per year (Ramakrishnan and Toky, 1981).

Many soil carbon-conserving practices are currently being implemented for reasons other than CO_2

TABLE 1.4 Assumed Performance of Various Hydrogen Production Methods

	Energy Input-Output Ratio	Levelized Cost for O&M and Capital ($/GJ H_2 produced)
Coal[a]	1.292	4.66
Natural gas[a]	1.115	1.72
Oil[b]	1.2	3.0
Biomass[a]	1.3	4.44
Electrolysis[a]	1.1	2.36

[a]Williams (1995).
[b]Kaarstad and Audus (1997).

NOTE: O&M=operation and maintenance.

mitigation, such as to control erosion or to reduce production costs or labor. However, impediments to more widespread adoption include costs of initial investment in equipment, fear or reluctance to adopt new technologies, and increased risks during the adoption transition phase. Financial incentives could substantially increase the rate of adoption of carbon-sequestering practices and potentially provide a significant addition to farm income. We have not developed a cost of emissions mitigation schedule for these technologies. In many circumstances, these technologies can be expected to penetrate the market even in a reference scenario since their adoption yields a net benefit. Clearly, further work is required. In addition, we ignore the problem of institutional mechanisms to ensure that the required change in agricultural practices is actually carried out. The purpose of this exercise is to examine the potential contribution of soil carbon capture to an overall strategy of global carbon management.

For the purposes of this analysis, we assume that 40 to 80 PgC are offset over the course of the next century. Given that the rate of carbon uptake by soils is highly nonlinear,[13] we assume that the rate of uptake is four times greater in the initial year of activity than 100 years later.

Aforestation and Reforestation

The forestry sector is modeled explicitly in MiniCAM. As the extent of forested lands expands and contracts, the associated stocks of carbon change endogenously with implied fluxes. When carbon is valued, land use changes endogenously. Carbon stocks in forests tend to accumulate, other things being equal; however, the model does not accumulate carbon in forests indefinitely. Rather, these stocks are harvested and employed as commercial biomass to offset fossil fuel use.[14]

Geologic Sequestration of CO_2

Several types of reservoirs could be employed for long-term storage of CO_2. The most obvious are oil and gas wells—both active and depleted—from which much of the CO_2 originated in the form of fossil fuel, but this reservoir is limited in capacity. Larger reservoirs are also potentially available, including saline geological formations, caverns, salt domes, and deep oceans. Herzog et al. (1997) have estimated the capacity of various reservoirs (Table 1.5).

Of these, oil and gas reservoirs are the most attractive. Active oil wells can utilize the carbon dioxide in tertiary recovery, and CO_2 can be used as a sweep gas to promote the production of methane from certain types of coal deposits. Furthermore, because these geological formations have already demonstrated their ability to hold oil and gas, there are no obvious environmental consequences to long-term storage in such locations. The same cannot be said for deep oceans and deep saline geological formations. Those sites, although promising, hold potential uncertainties in retention and environmental impacts.

Ocean disposal presents special problems. In addition to the problem of verification, oceans are only a partial reservoir. In recalling the earlier discussion of the carbon cycle, introducing CO_2 into the ocean through dispersion implies that a fraction of this quantity is committed to the atmosphere as well. The

---

[13]This nonlinear behavior is the consequence of the uptake rate of soils. While initially high, it declines with time as soil carbon content approaches the equilibrium, preindustrial level. We ignore potential interactions with atmospheric CO_2 concentrations.

[14]Forest products can also be used in a variety of applications that prevent or delay their oxidization, such as building materials, railroad ties, and telephone poles. However, the scale of forest products potentially available under a program that removed significant quantities of carbon from the atmosphere would quickly lower the price of these products to zero.

TABLE 1.5 Estimates of Global Carbon Storage Reservoirs

Carbon Storage Reservoir	Range (PgC)	
	Low	High
Deep ocean	1,391	27,000
Deep saline geological formations	87	2,727
Depleted gas reservoirs	136	300
Depleted oil reservoirs	41	191

ocean can be a permanent reservoir only if mechanisms can be found to isolate the carbon from the larger patterns of ocean circulation. The quantities of carbon captured over the course of the next century in the carbon-constrained scenarios examined here thus do not strain the potential capacity of total reservoirs.

However, while we recognize the potential of carbon capture and sequestration technologies to add an important new element to the set of climate mitigation technologies, it is also important to point out that these are part of a larger suite of technologies whose composition varies over time and space. The deployment of geological carbon capture and sequestration technologies may be limited by physical or technical considerations. It may also be limited by public acceptance, either locally or in general. Research can begin to address technical concerns, but experience will ultimately be needed to determine the acceptability of these technologies.

Sequestration of Carbon in Solid Form

None of the problems associated with sequestration are encountered if carbon is removed in the form of a solid. Lackner et al. (1995, 1997, 1999), for example, have investigated technologies that render the carbon as rock. As such it is highly stable, unlikely to return to enter the atmosphere, and for all practical purposes permanent.

Steinberg (1983, 1991, 1996,) and Steinberg and Grohse (1989) have examined processes that remove carbon in elemental form. Unlike the Lackner processes that react the carbon to rock, the Steinberg process leaves the carbon in a form that could potentially react if it is not properly disposed. However, if returned to depleted coal mines or other remote sites, the carbon could be monitored easily and verification of its location would be simple.

We have not explored these technologies in this analysis. Both are more expensive than the alternatives described earlier. However, they offer attractive monitoring and verification characteristics that may ultimately prove crucial.

FILLING THE EMISSIONS GAP

As noted earlier, stabilizing the concentration of carbon in the atmosphere requires global net carbon emissions to the atmosphere to peak then decline. In the model, this is accomplished by imposing a tax on the emission of carbon, which in turn induces technology substitutions relative to the reference case. A variety of policy instruments are available to society to signal the value of carbon. A tax is chosen because it leads to an economically efficient (least-cost) solution. However, without some signal from society that cumulative emissions will be limited, some variant of the reference case will emerge. In contrast to the suite of technologies that deploy in either the AOG or the CBF reference case, a wide range of low- and non-carbon-emitting energy technologies deploy more aggressively when CO_2 concentrations are limited.

One way to depict the role of technologies that become more important in an efficient emissions limitation regime is to associate the carbon emissions reductions with the appropriate technology. This is done for a 550-ppmv CO_2 concentration limit against a background of AOG and CBF in Figure 1.9. In this figure, emissions reductions are measured as the avoided carbon release associated with increased deployment of each technology. For example, as the carbon tax rises, more and more commercial biomass comes onto the market relative to the reference case and displaces carbon-emitting fuels (relative to the reference case). Thus, as time passes an ever-widening biomass wedge fills part of the gap. If a non- or low-carbon-emitting technology does not increase its absolute deployment as the value of carbon increases, no contribution is recorded by this method. As we show this approach can be extremely illuminating in some instances but can underrepresent the contribution of a technology in other instances.

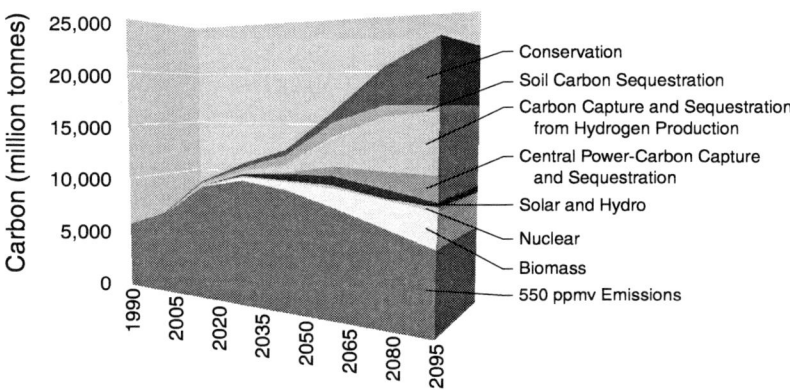

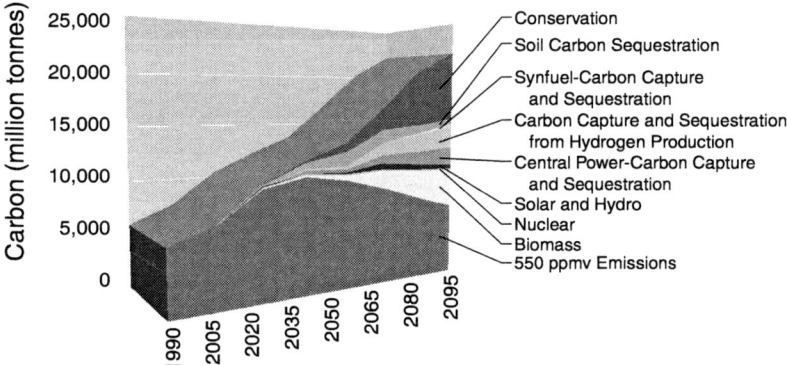

FIGURE 1.9 Comparison of technologies that fill the gap under alternative reference scenarios and a 550-ppmv CO_2 limit. Abscissa in millions of tonnes of emitted carbon displaced by conservation, sequestration, and alternate energy-producing technologies.

Several observations are worth making about these cases. We begin with the AOG case. First, technologies such as carbon capture and sequestration, commercial biomass, and hydrogen fuel cells that fill the gap are, with the exception of energy conservation, not presently significant contributors to the global energy system. Carbon capture and sequestration technologies, which expand dramatically, deploy only when carbon is valued. They do not deploy in a reference case. The presence of these technologies as a package reduces the minimum cost of achieving a 550-ppmv concentration limit by more than a factor of two.

Second, the technology portfolio changes over time. For example, soil carbon capture and sequestration are relatively important early on due to their low cost. Over time, their role declines due to both the exhaustion of the activity's potential and the rapid expansion in magnitude of the carbon emission mitigation requirement.

Third, some technologies are more important when others are also available. Hydrogen fuel cells are used extensively in conjunction with the capture and sequestration of carbon from the natural gas feedstock being used to obtain hydrogen. Natural gas, CH_4, is inexpensive in the AOG scenario. Since natural gas is mostly hydrogen, it is relatively easy and inexpensive to remove the carbon and sell the hydrogen. Without the ability to sequester the carbon and keep it out of the atmosphere, fuel cells would be far less effective in filling the gap.

Fourth, some technologies expand their relative importance without expanding their absolute deployment. Solar and nuclear power are examples of technologies whose relative importance increases, but whose absolute deployments are roughly comparable in both the reference case and the carbon concentration limit case. Both dramatically expand their share of power generation capacity between the reference and the 550-ppmv-ceiling case. (In 2095, non-carbon-emitting power generation is 50% of output in the reference case, but it grows to 80% in the 550-ppmv-ceiling case.) This increase in relative importance without expanded absolute deployment is the consequence of the interaction of conservation and power generation technologies. The expanded deployment of energy conservation leads to a reduction in the rate of growth of electricity demand. With a lower rate of expansion, the shift in technology from fossil fuel power generation toward non-fossil fuel power leaves the rate of growth of non-fossil fuel power technologies approximately the same as in both the reference and the CO_2 limit cases. As a consequence, the gap figure does not appropriately credit the contribution of solar and nuclear technologies to stabilizing the concentration of CO_2. That understanding requires a deeper analysis.

Regional Diversity

The world is not a homogeneous place, and technology deployment varies strongly across regions and over time. Regional analysis indicates China, for example, has a relatively smaller expansion than India in the deployment of commercial biomass technology. Differences in technology expansion rates and emphasis can be traced to the multitude of resource, cultural, institutional, and policy environments that exist around the world.

Effect of Alternative Reference Technology Backgrounds

The paths of economic and technological development are difficult to predict. Contrast the CBF and AOG cases, for example. They differ by only one assumption—that lower grades of more abundant oil and gas become available at approximately current prices. This one assumption drives important differences in the technologies that are deployed to limit CO_2 emissions.

If we focus on the change between the reference case and the carbon-limited case, we see both similarities and differences between CBF and AOG in Figure 1.9. In both CBF and AOG, soil carbon plays an important role early in the twenty-first century, but the role of carbon capture and sequestration from hydrogen production is greatly decreased in the CBF case compared to AOG. In contrast, the direct effects of energy conservation induced by the carbon tax are smaller in the CBF case than in AOG. After all, because cheap oil and gas are not available in the CBF reference case, end-use energy is already more expensive, and many conservation measures have already been implemented.

However, the indirect effects of energy conservation are significantly greater. Since much of the end-use energy in the CBF case is either electricity or synthetic liquids and gases derived from coal, there is an energy conversion penalty. Every joule of energy conserved at the point of use conserves 1 to 2 joules of the original fossil fuel resources—primary energy. Thus, significant reductions in emissions are associated with a smaller synfuels industry, which in turn is the direct consequence of increased energy conservation.

Commercial biomass production plays a more important role in both the AOG and the CBF cases. The higher energy prices imply greater profitability for commercial biomass, but the expansion is greater in the CBF case.

RISK MANAGEMENT

Stabilizing the concentration of greenhouse gases requires that the global energy system undergo radical change over the course of the twenty-first century. Reference energy systems by the end of the twenty-first century are dramatically different energy systems that stabilize the concentration of atmospheric CO_2. The nature and composition of future energy systems are subject to many uncertainties.

Critical uncertainties include reference technology evolution (how good will the reference technologies turn out to be?), policy uncertainty (what concentration will ultimately be chosen, either explicitly or implicitly, as the concentration limit?), institutional uncertainty (what institutional mechanisms will be utilized to limit cumulative emissions—cap and trade, taxes, regulations?), regional diversity (what technologies will be needed in different regions around the world?), and investment uncertainties (which investments in R&D will be made, which will succeed, and which will fail?).

If the future technology evolution were known, and a technology portfolio were to be deployed rationally over space and time, the problem of framing a set of technology R&D investments would be difficult enough. Yet even under these relatively simple circumstances, our results show that no single technology dominates society's response to climate change. A regionally and temporally variable mix of technologies will emerge to form the global response to climate change. The existence of a multitude of additional uncertainties, only some of which are subject to human influence, greatly complicates the selection of an R&D portfolio that lays the foundation for a effective technological response to carbon management.

Only investments in energy R&D can create a technology portfolio to address a cumulative emissions limit, and although the results of R&D investments are uncertain—it is impossible to predict the benefit of each dollar invested—successful investments yield large social benefits. Figure 1.10 shows the reduction in minimum, present discounted cost of satisfying alternative concentration limits as solar photovoltaic technology improves. Depending on the reduction in cost and the concentration limit, the present discounted minimum cost reduction ranges from tens of billions of dollars to trillions of dollars.

The cumulative nature of the relationship between emissions and concentrations of CO_2, coupled with the nature of technology transitions, dictates that investments in energy R&D are required in the near term. Energy systems take from 50 to 100 years for a modal change (see, for example, Marchetti, 1979; Häfele, 1981; Nakicenovic et al. 1998). For example, Figure 1.11 shows the penetration of the

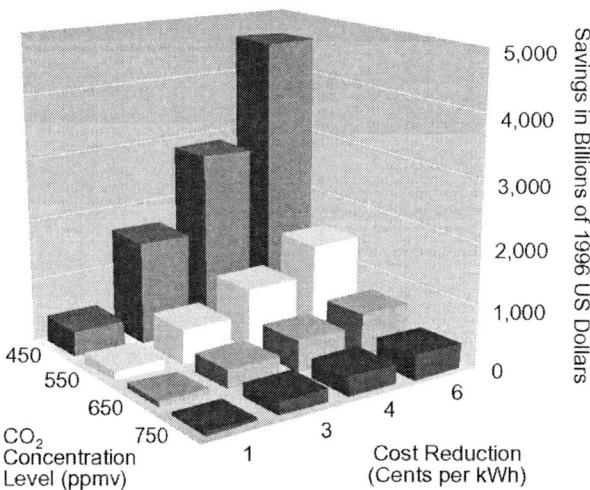

FIGURE 1.10 The value of improving solar electric power technologies relative to a reference case cost of 8 cents per kilowatt-hour in 2020.

automobile into the French transportation system. It took half a century for the now-dominant technology to capture half of the market. Market penetration on a global scale takes longer.

Table 1.1 indicates that depending on the limit placed on the atmospheric concentration of CO_2, the date at which *global* emissions peak ranges from 2005 to 2062. This in turn implies that beyond those "peak" dates, new investments in capital stocks to provide energy services must be predominantly non-carbon emitting. That is, additions to the energy-related capital stock must improve end-use energy efficiency, provide energy with non-fossil energy forms, or provide a mechanism by which to capture and sequester carbon.

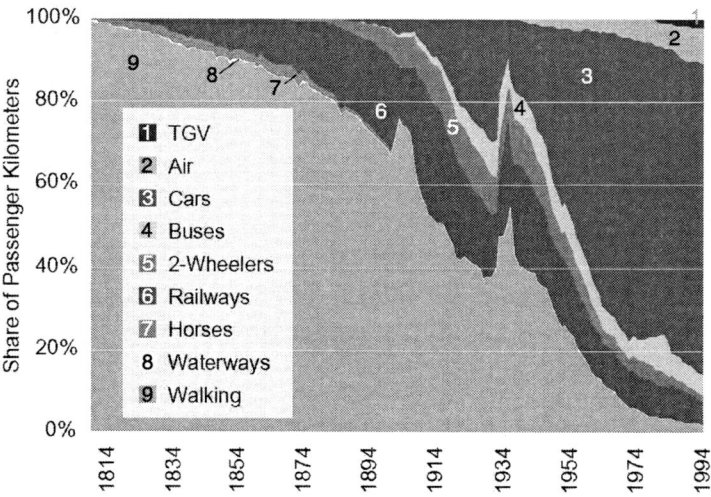

FIGURE 1.11 The changing composition of French transport since 1800. NOTE: (TGV = modern high-speed trains).

A constraint on the concentration of CO_2 therefore carries implications for energy R&D. First, it must be broad based. Multiple energy technologies play important roles in limiting cumulative emissions of carbon. No single technology ever captures more than a fraction of the global energy system market. Furthermore, it is impossible to know a priori even the relative importance of different technologies. The multitude of uncertainties that characterize a century-scale analysis provides an infinite set of potential combinations of technology deployments.

Fortunately, decisions need not be taken today that would remain in effect over the next century. Rather, decisions taken today shape the options available at the next decision point. As time proceeds, knowledge grows. This growth in knowledge will ultimately lead to "technology maturation" in some areas but will also suggest new potentially useful R&D investments in others. This is the principle of "act-then-learn-then-act."

GLOBAL ENERGY R&D

We have found that the long-term value of energy technology improvement is high. However, current energy R&D is declining, is not coordinated internationally, and has yet to significantly incorporate a climate change motivation.

Nine nations account for more than 96% of the industrialized world's public-sector energy R&D investments. Examining trends within the energy R&D programs of these nations (Figure 1.12) can provide some insight into the world's commitment to energy R&D.

The United States and Japan carry out about 80% of current energy R&D. This is a dramatic shift since 1985, when Europe accounted for more than a third of the total. Energy R&D programs carried out under the European Union's Framework Program have also declined 10% in real terms over this time period.

Comparing these data is complicated by the problem of exchange rate valuation. Using purchasing power parity exchange rates shows that although the composition of Figure 1.12 changes, the qualitative

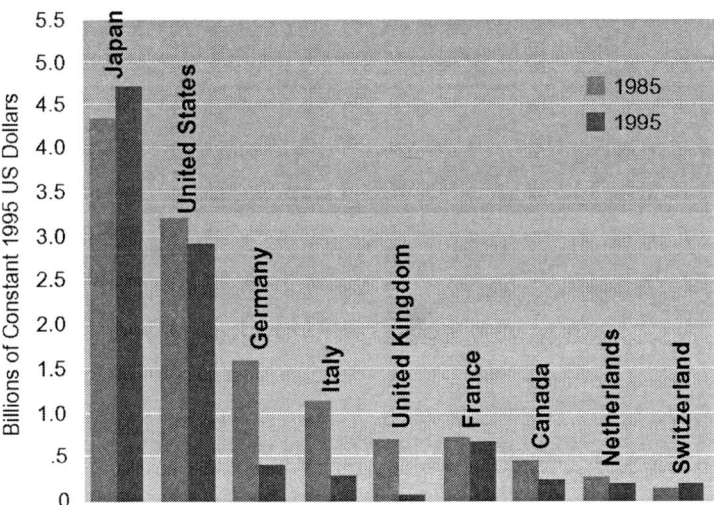

FIGURE 1.12 Public-sector energy R&D expenditures, 1985 and 1995 (millions of constant 1995 U.S. dollars).

TABLE 1.6 Public-Sector Energy R&D (purchasing power parity exchange rates in dollars)

Year	United States	Japan	European Union	Germany	United Kingdom	Canada	Total
1985	3,269	2,366	749	1,470	424		8,279
1986	3,052	2,433	515	1,168	424		7,592
1987	2,658	2,106	281	866	424		6,336
1988	2,657	2,217	518	801	371		6,565
1989	2,902	2,285	425	736	357		6,705
1990	3,427	2,185	317	737	299	195	7,161
1991	3,255	2,216	378	679	257	210	6,995
1992	3,335	2,234	364	602	219	205	6,960
1993	2,807	2,288	275	523	195	205	6,293
1994	2,944	2,449	355	441	151	199	6,539
1995	2,856	2,532	412	425	87	201	6,514
1996	2,465	2,621	497	431	76	195	6,285
1997	2,160	2,511	485	394	58	254	5,862
1998	2,084	2,394	474	396	51	183	5,582
% Decline, 1985-1998	36	−1	37	73	88	6 (1990-1998)	33

SOURCE: Dooley and Runci (2000).

character does not. In fact, the trend toward lower investment in energy R&D has, if anything, accelerated in some nations (Table 1.6). Only in Japan do energy R&D budgets rise over the period since 1998.

The general decline in global energy R&D is not restricted to the public sector. Where data are available for the private sector, a similar pattern is observed. For example, the United States experienced an even steeper decline in private sector energy R&D than in public-sector funding. From 1985 to 1996, public-sector funding in the United States declined by 25%. Private-sector energy R&D declined by more than 40%.

The decline in energy R&D is not particularly surprising. Since 1981, the real price of oil has decreased in the United States. In 1986, the oil price fell precipitously. In an environment in which the scarcity value of energy has declined, the expected return on investments in energy R&D declines as well. More recently, the electric power sector has experienced dramatic institutional changes. Deregulation has proceeded on many fronts, with Europe leading the way. The deregulation of power generation changed the incentive structure for the power sector by shifting emphasis away from long-term considerations toward short-term profitability. Energy R&D investments, the preparation for the long term, have declined accordingly.

The decline in investments might be considered an appropriate reflection of the energy situation were it not for the fact that it does not incorporate a serious consideration of the value of energy R&D in meeting long-term climate goals. If limits were expected for the concentration of CO_2, then the value of energy R&D would have to rise accordingly.

Technologies whose deployment expands dramatically when the concentration of CO_2 is limited are of particular interest. Figure 1.13 shows energy R&D in two groups—technologies that are of broad general importance regardless of whether or not climate change is an issue and technologies for which climate change can be extremely important in shaping their deployment.

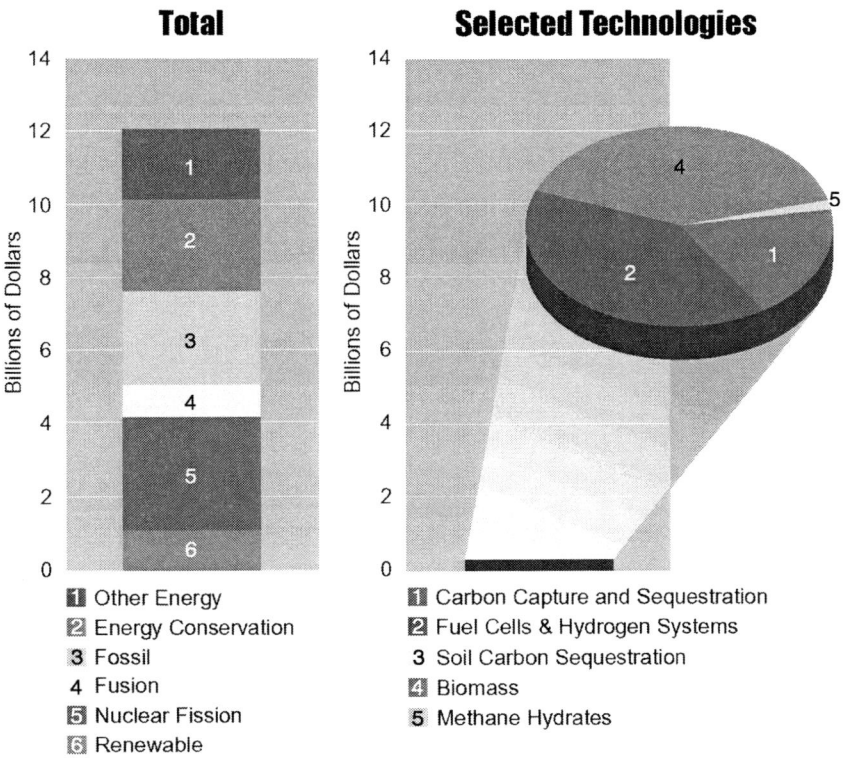

FIGURE 1.13 Global energy R&D expenditures in 1995 (billions of constant 1995 U.S. dollars).

Energy R&D is no more homogeneous than the energy system it serves. It is composed of investments undertaken by both public- and private-sector agents. It includes investments that range from basic energy-related sciences to technology deployment. Building an R&D portfolio requires the inclusion of investments across this full spectrum. The most difficult components of the portfolio are those that occur at the interface of the public and private sectors. Here there are more numerous lessons in how *not* to proceed than in how to proceed. The potential for "technology pork" is an ever-present danger (Cohen and Noll, 1991). More important than wasted resources represented by political R&D are the wasted opportunities to prepare useful options for the future. This is particularly true in light of the long-term, sustained nature of the carbon management problem.

THE CHALLENGE

Climate change constitutes one of the principal challenges of the twenty-first century, and one of the major dimensions of this challenge is technological. Net human carbon dioxide emissions to the ocean-atmosphere must eventually decline to zero for its concentration to be stabilized. This follows from the fact that long-term, steady-state concentrations are determined by cumulative emissions, not by the emissions rate. As such, CO_2 is fundamentally different from other greenhouse gases. This feature of the climate problem has several important, near-term implications for the character of efficient policy development.

Atmospheric stabilization requires fundamental change in the global energy system. The present global energy system relies primarily on fossil fuels. Reliance on fossil fuels grew steadily during the preceding century, and there are sufficient economic fossil fuel resources to enable this trend to continue throughout the twenty-first century. The continued spread of economic development will ensure a continued expansion in the demand for energy services. To stabilize the concentration of CO_2 requires the development of cost-effective, low- and non-carbon-emitting energy technologies. In short, an energy technology revolution is needed unlike any in history. The specter of climate change can be avoided by limiting the atmospheric concentration of carbon dioxide, but this will be possible only by developing technology that will change the way in which the world uses low-cost fossil fuel resources, with an associated increase in the cost of energy. This is not the usual motivation of private-sector technology development.

This revolution will not happen by magic. It will occur only because investments were made to develop the technologies. Public policy will play an important role both in signaling the need for new technologies and in facilitating their development and deployment.

Investments will be needed that span the full spectrum from basic energy-related sciences to technology demonstration, and it will be important to reverse the present decline in global energy R&D. Yet, given that a new global energy system is needed to address a global problem, it will also be necessary to coordinate the R&D efforts of many nations and of both the public and the private sectors. Contributions will be needed from both the public and the private sectors. Perhaps the most daunting challenge will be to adequately incorporate public-sector motivation for carbon management into private-sector R&D decision making.

Finally, while a strategy to develop and deploy technology will be essential to a program to minimize the cost of addressing climate change, it is just part of a larger portfolio. That larger portfolio includes scientific research, emissions limitation, and adaptation to climate change. As progress is made in these areas, the R&D portfolio must also be able to evolve. Given the complexity and time frame of the climate problem, flexibility is of the essence.

ACKNOWLEDGMENTS

The work reported here is an outgrowth of research conducted under the Battelle Global Energy Technology Strategy Project to Address Climate Change (GTSP). That work is sponsored by both public- and private-sector organizations, including Battelle Memorial Institute, BP, the Electric Power Research Institute (EPRI), ExxonMobil, Kansai Electric Power, National Institute for Environmental Studies (Japan), New Energy and Industrial Technology Development Organization (Japan), North American Free Trade Agreement Commission for Environmental Cooperation, PEMEX (Mexico), Tokyo Electric Power, Toyota Motor Company, and the U.S. Department of Energy.

The project is guided by a Steering Group, whose members have roots in the public and private sectors as well as academics and the nongovernmental organization communities. Members include Richard Balzhiser, president emeritus, EPRI, Richard Benedick, former U.S. ambassador to the Montreal Protocol; Ralph Cavanagh, co-director, Energy Program, Natural Resources Defense Council; Charles Curtis, executive vice president, United Nations Foundation; Zhou Dadi, director, China Energy Research Institute; E. Linn Draper, chairman, president and chief executive officer, American Electric Power; Daniel Dudek, senior economist, Environmental Defense Fund; John H. Gibbons, former director, Office of Science and Technology Policy, Executive Office of the President; José Goldemberg, former environment minister, Brazil; Jim Katzer, strategic planning and programs manager, ExxonMobil; Yoichi Kaya, director, Research Institute of Innovative Technology for the Earth, Government of Japan; Robert McNamara, former president, World Bank; Hazel O'Leary, former secretary, U.S. Department of Energy; Rajendra K. Pachauri, director, Tata Energy Research Institute; Thomas

Schelling, distinguished university professor of economics, University of Maryland; Hans-Joachim Schellnhuber, director, Potsdam Institute for Climate Impact Research; Pryadarshi R. Shukla, professor, Indian Institute of Management; Kathryn Shanks, vice president, Health, Safety and Environment, BP; Gerald Stokes, assistant laboratory director, Pacific Northwest National Laboratory; John Weyant, director, Stanford Energy Modeling Forum; and Robert White, former president, National Academy of Engineering.

Although the research was lead by the scientists at Battelle, the project benefited from the work of the following collaborating institutions: Autonomous National University of Mexico, Centre International de Recherche sur l'Environnement et le Developpement (France), China Energy Research Institute, Council on Agricultural Science and Technology, Council on Energy and Environment (Korea), Council on Foreign Relations, Indian Institute of Management, International Institute for Applied Systems Analysis (Austria), Japan Science and Technology Corporation, National Renewable Energy Laboratory, Potsdam Institute for Climate Impact Research (Germany), Stanford China Project, Stanford Energy Modeling Forum, and the Tata Energy Research Institute (India).

Without the help and support of the above individuals and organizations, the work reported here would have been impossible. The authors are most grateful for all of their many contributions.

REFERENCES

Adams, D., P. Freund, P. Reimer and A. Smith. 1996. Technical response to climate change. International Energy Agency Greenhouse Gas Program.

Benson, S., W. Chandler, J. Edmonds, M. Levine, L. Bates, H. Chum, J. Dooley, D. Grether, J. Houghton, J. Logan, G. Wiltsee, and L. Wright. 1998. *Carbon Management: Assessment of Fundamental Research Needs.* Office of Energy Research, U.S. Department of Energy. DOE/ER-0724. Washington, D.C. August 1997.

Brown, R.A., M.J. Rosenberg, W.E. Easterling III, and C. Hays. 1998. *Potential Production of Switchgrass and Traditional Crops Under Current and Greenhouse-Altered Climate in the MINK Region of the Central United States.* Report No. PNWD-2432. Pacific Northwest National Laboratory, Washington, D.C.

Cohen, L.R., and R.G. Noll. 1991. *The Technology Pork Barrel.* Brookings Institution, Washington, D.C.

Cole, V., C. Cerri, K. Minami, A. Mosier, N. Rosenberg, and D. Sauerbeck. 1996. Agricultural options for mitigation of greenhouse gas emissions. Pp. 744-771. In: R.T. Watson, M.C. Zinyowera, and R.H. Moss (eds.) *Climate Change 1995: Impacts, Adaptations, and Mitigation of Climate Change: Scientific-Technical Analyses.* Contribution of Working Group II to the Second Assessment Report of the Intergovernmental Panel on Climate Change. Cambridge University Press, Cambridge, U.K.

Dooley, J. 1998. Unintended consequences: energy R&D in a deregulated energy market. *Energy Policy* 26(7):547-555.

Dooley, J.J. and J.A. Edmonds. 1997. *Workshop for the Preliminary Identification of Basic Science Needs and Opportunities for the Safe and Economical Capture and Sequestration of CO2.* Published in Bensen, S., W. Chandler, J. Edmonds, M. Levine, L. Bates, H. Chum, J Dooley, D. Grether, J. Houghton, J. Logan, G. Wiltsee, and L. Wright. 1999. B. Eliasson, P. Riemer, and A. Wokaun (eds.) Greenhouse gas control technologies. Pergamon Press, New York. Pp 457-462.

Dooley, J.J. and P.J. Runci. 2000. Developing nations, energy R&D, and the provision of a planetary public good: a long-term strategy for addressing climate change. *Journal of Environment and Development.* V. 9(3):216-240.

Edmonds, J. and J. Reilly. 1985. *Global Energy: Assessing the Future.* Oxford University Press, New York.

Edmonds, J.A., J.M. Reilly, R.H. Gardner, and A. Brenkert. 1986. *Uncertainty in Future Global Energy Use and Fossil Fuel CO2 Emissions 1975 to 2075.* TR036, DO3/NBB-0081 Dist. Category UC-11. National Technical Information Service, U.S. Department of Commerce, Springfield, Va.

Edmonds, J., M. Wise, R. Sands, R. Brown, and H. Kheshgi. 1996. *Agriculture, Land-Use, and Commercial Biomass Energy: A Preliminary Integrated Analysis of the Potential Role of Biomass Energy for Reducing Future Greenhouse Related Emissions.* PNNL-11155. Pacific Northwest National Laboratories, Washington, D.C.

Edmonds, J., M. Wise, H. Pitcher, R. Richels, T. Wigley, and C. MacCracken. 1997. An integrated assessment of climate change and the accelerated introduction of advanced energy technologies: an application of MiniCAM 1.0. *Mitigation and Adaptation Strategies for Global Change* 1:311-339.

Freund, P., and W.G. Ormerod. 1997. Progress toward storage of carbon dioxide. *Energy Conversion and Management* 38:S199-S204.

Gebhart, D.L., H.B. Johnson, H.S. Mayeux, and H.W. Pauley. 1994. The CRP increases soil organic carbon. *Soil and Water Conservation* 49:488-492.

Gottlicher, G., and R. Pruschek. 1997. Comparison of CO_2 removal systems for fossil-fuelled power plant processes. *Energy Conversion and Management.* 38:S173-S178.

Grether, D., J. Houghton, J. Logan, G. Wiltsee, and L. Wright. *Carbon Management: Assessment of Fundamental Research Needs.* U.S. Department of Energy, Office of Energy Research, Washington, D.C.

Häfele, W. 1981. *Energy in a Finite World.* Ballinger Publishing Company, Cambridge Mass.

Herzog, H., E. Drake, and E. Adams, 1997. CO_2 capture, reuse, and storage technologies for mitigating global climate change. DOE order number DE-AF22-96PC01257.

IPCC (Intergovernmental Panel on Climate Change). 1996a. *Climate Change 1995: The Science of Climate Change. The Contribution of Working Group I to the Second Assessment Report of the Intergovernmental Panel on Climate Change.* Cambridge University Press, Cambridge, U.K.

IPCC (Intergovernmental Panel on Climate Change). 1996b. *Climate Change 1995: Impacts, Adaptation, and Mitigation of Climate Change: Scientific-Technical Analysis. In the Contribution of Working Group II to the Second Assessment Report of the Intergovernmental Panel on Climate Change.* Cambridge University Press, Cambridge, U.K.

IPPC (Intergovernmental Panel on Climate Change), 1995. *Climate Change 1994: Radiative Forcing of Climate Change and An Evaluation of the IPCC IS92 Emissions Scenarios.* J.T. Houghton, L.G.M. Filho, J. Bruce, H. Lee, B.A. Callander, E. Haites, N. Harris, and K. Maskell (eds.), Cambridge University Press, Cambridge, United Kingdom.

Jenkinson, D.S. 1971. The accumulation of organic matter in soil left uncultivated. *Rep. Rothamsted Exp. Stn. for 1970* (Part 2):113-117.

Kaarstad, O., and H. Audus. 1997. Hydrogen and electricity from decarbonized fossil fuels. *Energy Conversion and Management* 38:S431-S436.

Kaku, M. 1997. *Visions: How Science Will Revolutionize the 21st Century.* Anchor Books, N.Y.

Kinzey, B.R., M.A. Wise, and J.J. Dooley. 1998. Fuel cells: a competitive market tool for carbon reductions. *Proceedings of the 21st Annual Conference of the International Association of Energy Economics.* Quebec, Canada.

Lackner, K.S., C.H. Wendt, D.P. Butt, E.L. Joyce, and D.H. Sharp. 1995. Carbon dioxide disposal in carbonate minerals. *Energy* 20:1153-1170.

Lackner, K.S., D.P. Butt, and C.H. Wendt. 1997. Magnesite disposal of carbon dioxide. Pp. 419-430 in *Proceedings of the 22nd International Technical Conference on Coal Utilization and Fuel Systems*, Clearwater, Fl.

Lackner, K.S., H.J. Ziock, and P. Grimes. 1999. Carbon dioxide extraction from air: is it an option? Pp. 885-896 in *Proceedings of the 24th International Technical Conference on Coal Utilization and Fuel Systems*, Clearwater, FL.

Manne, A.S., and R. Richels. 1997. *Toward the Stabilization of CO2 Concentrations—Cost-Effective Reduction Strategies.* Stanford University, Stanford, Calif.

Marchetti, C. 1979. Energy systems: the broader context. *Technological Forecasting and Social Change* 15:79-86.

Nakicenovic, N., A Grübler, and A. McDonald (eds.). 1998. *Global Energy Perspectives.* Cambridge University Press, Cambridge, U.K.

Nakicenovic, N., et al. 2000. *Special Report on Emissions Scenarios.* Cambridge University Press, Cambridge, U.K.

Pepper, W., et al. 1992. Emission Scenarios for the IPCC, An Update. Assumptions, Methodology, and Results. IPCC Special Report on Emission Scenarios.

Ramakrishnan, P.S., and O. P. Toky. 1981. Soil nutrient status of hill-agroecosystems and recovery after slash and burn agriculture (Jhum) in north-eastern India. *Plant and Soil* 60:41-64.

Richels, R., J. Edmonds, H. Gruenspecht, and T. Wigley. 1996. The Berlin Mandate: the design of cost-effective mitigation strategies. *Climate Change: Integrating Science, Economics and Policy,* CP-96-1:29-48. N. Nakicenovic, W.D. Nordhaus, R. Richels, and F.L. Toth (eds.). International Institute for Applied Systems Analysis, Laxenburg, Austria.

Rogner, H.-H. 1996. *An Assessment of World Hydrocarbon Resources.* WP-96-56. International Institute for Applied Systems Analysis, Laxenburg, Austria.

Shell Oil. 1995. *The Evolution of the World's Energy System 1860-1960.*

Steinberg, M. 1983. *An Analysis of Concepts for Controlling Atmospheric Carbon Dioxide.* TR007, DOE/CH/00016-1. Washington, D.C.

Steinberg, M. 1991. *Biomass and Hydrocarb Technology for Removal of Atmospheric CO2.* BNL-44410R (Rev. 2/91). Brookhaven National Laboratory, Upton, N.Y.

Steinberg, M. 1996. *Coprocessing Coal and Natural Gas for Liquid Fuel with Reduced Greenhouse Gas CO2 Emission.* Brookhaven National Laboratory, Upton, N.Y.

Steinberg M., and E.W. Grohse. 1989. *The Hydrocarb Process for Environmentally Acceptable and Economically Competitive Coal-Derived Fuel for the Utility and Heat Engine Market*. BNL-43554. Brookhaven National Laboratory, Upton, N.Y.

United Nations. 1992. *Framework Convention on Climate Change*. New York.

Wigley, T.M.L., R. Richels, and J. Edmonds. 1996. Economic and environmental choices in the stabilization of atmospheric CO_2 concentrations, *Nature* 379(6562):240-243.

Williams, R.H. 1995. *Variants of a Low CO2-Emitting Energy Supply System (LESS) for the World*. PNL-10851. Pacific Northwest Laboratory, Richland, Wash.

DISCUSSION

George Helz, University of Maryland: I just wanted you to clarify something related to Figure 1.1. I thought you said that to stabilize CO_2 in the atmosphere, emissions had to go to zero, which I would interpret to mean that all carbon oxidation has to stop, and I don't think that's what you really meant.

James Edmonds: No.

George Helz: Could you clarify what you meant by emissions?

James Edmonds: Let me suggest a really simple model to work with. Suppose you've got two buckets and you put all the carbon on the planet in two buckets. One is a terrestrial fossil fuel bucket, and the other is an ocean-atmosphere bucket. The net transfer from the terrestrial fossil fuel bucket into the ocean-atmosphere bucket has to go to zero, but that doesn't mean oxidation is going to zero. You've got a carbon cycle that has a lot of offsetting entries in it, but the net flow has to go to zero for the concentration to be stable.

Geraldine Cox, EUROTECH: I'm going to be politically incorrect and ask about one of your assumptions, and that is on population growth because it really wasn't explicit. To me, population growth or the challenge it presents is going to be the overriding fact in all of this.

James Edmonds: Thank you for bringing that up. You are absolutely right, demographics is going to be one of the absolutely critical things over the course of the century. We don't know how it will play out, but it is really interesting. We have been holding a series of workshops on demographics because we are convinced that you are absolutely right—that this is a central issue. It may be that what we have is a problem that, over the course of the century, looks very similar to the problem that we have today—relatively high fertility rates outside of Organization for Economic Cooperation and Development (OECD) countries and a growing population. Underlying that scenario is a world population of 11 billion.

Recently, most population numbers have been revised downwards. The best estimates are below 10 billion, reflecting the rapid decline in fertility rates around the world, in both developed and developing countries. So it reflects some success in a variety of different dimensions. The numbers have been declining. Tom Schelling at the University of Maryland suggested that the biggest problem facing us is keeping the rate of population growth from overwhelming the resource base. We may end the twenty-first century with the fertility rate dropping through the floor, while we try to maintain a population on the planet.

Furthermore, this is 100 years, and there is a revolution in the biological sciences. What does that revolution mean? Well, it may mean that if you are born in the middle of the century and you don't meet an end by stepping out in front of truck, or getting into an altercation with your neighbor, you may, for all practical purposes, live forever.

So you may have a really different demographic out there at the end of this century than anybody is thinking about. It is a really important question.

Alan Wolsky, Argonne National Laboratory: This is a very simple question. You spoke in terms of CO_2 concentrations, but I know that if you paint the window with three coats of black paint, the second and third coats aren't nearly as efficacious as the first coat. In a simple theory, the effect might be an exponential relationship. What is the "first coat" of CO_2? What's the concentration of CO_2 beyond which it doesn't really matter whether you add more CO_2 to the atmosphere?

James Edmonds: Are you asking when the temperature would increase to a point that it wouldn't matter anymore, or is the question one of when does it no longer pay to control and you ought to just let it ride?

Alan Wolsky: It's the simple first question.

James Edmonds: The artistry of the FCCC goal was that it said just don't do anything dangerous, and you're asking, I think, "What is dangerous?" That question has not been answered.

Alan Wolsky: Let me illustrate my question by asking another and giving a speculative answer. We know water is a greenhouse gas. We know water is given off when you burn methane. So why aren't people upset by increasing the amount of water in the atmosphere? I speculate that the correct answer is that there is already so much water in the atmosphere, that the likely anthropogenic increment doesn't matter.

James Edmonds: It's a simple question. It doesn't have a simple answer.

Alan Wolsky: Does it have an order-of-magnitude answer?

James Edmonds: Yes, if you say I think that if we ran most of the crop models at concentrations of 1,200 parts per million, you would have a hard time keeping agriculture operating. Nobody has actually done that, as far as I know, and it sounds like dinner conversation.

Klaus Lackner, Los Alamos National Laboratory: I'm really very much in agreement with you on the issue that we have to go down to zero in emissions. I do want to point out, on the other hand, the scales of the other carbon reservoirs and how long it would take to flood them. This is something you didn't point out, even though you have the numbers there. Vegetation contains roughly 600 gigatons of carbon, which corresponds to only about 100 years of fossil energy output at the current rate of emission.

Similarly, the ocean contains about 1,000-1,500 gigatons of carbon in the form of CO_2, and doubling this would change the pH from top to bottom by roughly 0.3. So overall, these other reservoirs, (ocean, biomass, and soil), which naturally are sinks, cannot really take up all of the carbon emitted. The scale at which we are generating CO_2 is very large compared to the size of the natural reservoirs.

James Edmonds: That's exactly right, except for the ocean, of course.

2

Carbon Dioxide Mitigation: A Challenge for the Twenty-First Century

David C. Thomas
BP Amoco Corporation

Life requires energy to survive. All living things consume fuel, generate the energy they need, and emit waste products. Our society is no different from the smallest one-cell organism in that we search for fuel and consume it to generate the energy we need to survive. The common denominator is that the primary fuel is carbon-based and the dominant waste is CO_2. As our society grows, its desire for more energy grows. Over the past 150 years, we have consumed enormous amounts of carbon-based fuels in developing our civilization. Increasingly, society has become concerned with the impact our actions are having on the planet and its ability to sustain our continued development.

The twentieth century was characterized by the development of an active environmental agenda that demonstrated society's concerns for clean air and water, minimizing waste and developing recycling, controlling chemical and radioactive emissions, and protecting endangered species. These concerns had dramatic effects on how we live and carry out our business affairs. As our understanding of this agenda matured, other concerns with an even greater potential for impact on the world have emerged. Climate change, deforestation, availability of plentiful potable water, biodiversity, and the interactions between them may be the defining environmental issues for the twenty-first century.

Climate change issues became an active topic for both scientific and political debate during the last decade. The recent Sixth Convention of the Parties of the United Nations Framework Convention on Climate Change (COP–6) meeting showed the intensity of concern and the range of viewpoints among the earth's nations. Some people counsel for mitigation action now, while others argue for more definite signs of climate change before taking action. The scientific community around the world is evaluating the validity of the climate change claims, critiquing approaches to mitigation, and developing mitigation options and strategies. The political community is developing equitable policies to share the burden of mitigation among the world's peoples. It needs the results from the scientific community as the basis for the decisions that will affect the lives and livelihoods of billions of people. The public, on whose behalf these activities are occurring, shows a wide range of understanding about climate change—extending from those who are largely unaware of the issue to those who are well informed.

The business community shows the same range of concern and understanding. Some are barely aware of the issue, whereas others have been involved in the discussion from the beginning. Many companies view the climate change issue as a threat to their existence and economic health; others see potential opportunities. All are concerned that mitigation will raise the cost of the goods and services they provide to the public, with negative consequences for all.

INDUSTRIAL GOALS

The prospect of global climate change is a genuine concern for the public and one that BP shares. The amount of CO_2 in the atmosphere is increasing and the temperature of the earth's surface is rising. Although there is uncertainty about the magnitude and consequences of these developments, the balance of informed opinion is that humans are having a discernible effect on the climate, and scientists believe that there is a link between the amount of CO_2 in the atmosphere and increased temperature. Faced with this uncertainty, BP believes that adopting a precautionary approach to climate change is the only sensible way forward in these circumstances. BP proposes to make real, sustainable, and measurable changes in its business practices. This is why BP has set for itself a voluntary goal to reduce its direct, equity share emissions of greenhouse gases by 10% from a 1990 baseline by 2010.

BP is active in a broad range of climate change issues in both the policy and the scientific arenas.[1] BP engineers and scientists study the effects our businesses have on greenhouse gases, water management, and biodiversity. We sponsor internal and external research that adds to both our understanding of business opportunities and our scientific understanding of environmental issues. BP Solar is the largest manufacturer and marketer of photovoltaic devices for producing electricity from solar radiation.[2] BP Energy provides energy management services to diverse businesses worldwide. BP recently sponsored the Hydrogen Interactive—First Contact as a way to introduce our interest in hydrogen as a fuel source and provide a forum for discussion and debate about the hydrogen economy.[3] BP's refining and marketing arm is developing clean fuel technology and innovative marketing concepts that introduce and showcase technologies to the consumer.[4] We have worked with Environmental Defense to develop an emission-trading methodology and market within the group as a learning and implementation tool.[5] This list of actions and activities is not exhaustive but gives a flavor of the voluntary actions that BP is using to support our group's commitment to its environmental responsibilities.

BP's reduction goals are even more aggressive when projected business growth is considered. The targeted 10% reduction below our 1990 baseline translates into a real reduction of more than 30% in projected 2010 CO_2 emissions. It is even more daunting when one considers that there are no economic incentives outside our internal goals. Many of our business activities involve partnerships with other companies that do not share our specific goals. Many of our partners are working through the process to develop targets that make sense for their companies, and we are gratified by the positive reception that our goals have received. Our discussions suggest that many companies are going through the same evaluative process that BP began in the 1990s.

In Chapter 1, Jae Edmonds presents predictions of climate change over the next several centuries.[6] This chapter discusses the approach that BP has taken to address the growing worldwide concern about carbon management or CO_2 mitigation.

CO_2 mitigation as a way to reduce man's impact on the environment is in its earliest stages of development. Economic incentives to mitigate CO_2 are rare and must be developed. These incentives can come through revenue streams from useful products, savings from the CO_2 mitigation activity, taxation, or government-sponsored emission-trading programs.

Companies cannot remain viable if they disadvantage themselves economically with respect to their competitors. Present CO_2 management projects concentrate on emission reductions from current operations through energy efficiency improvements, operating practice changes, or process changes. These actions must compete with other investment opportunities within the company and must generate a favorable return on the capital or operating funds invested.

Figure 2.1 provides a schematic view of the relative capital investments that will be needed to mitigate large amounts of CO_2. Mitigation costs will increase as the simpler options are completed. It will be necessary to reduce the costs of new mitigation technologies to ensure implementation. Our overall goal is to extend the low-cost curve while reducing the slope of the curve for higher-cost options. BP is active in addressing each of the technical directions indicated.

BP's CO_2 management strategy can be stated simply as follows:

• *Reduce energy consumption from manufacturing our products, develop cost-effective separation technologies for the CO_2 we do emit, find ways to use CO_2 in beneficial ways, and finally store any remaining CO_2 safely.* Figure 2.2 shows the overall approach schematically. In addition to implementing CO_2 reduction options now, we are also developing needed technology improvements in separation and storage.

• *Lead development of a viable CO_2 emission trading credit system to ensure that the lowest-cost options for abatement of CO_2 are found and implemented.* BP's approach to this important option is described on its Web page and is not discussed further in this chapter.[7]

CO_2 EMISSION REDUCTION

Emissions reduction is the starting point for any mitigation program. This involves taking a fresh look at ongoing operations to challenge standard operating practices from the new perspective of reducing CO_2 emissions. Each plant or refinery strives to optimize its operations for efficient production of the desired product. In the past, this process has not specifically included reduction in CO_2 emissions. BP has reviewed many of its operations and continues to do so through an ongoing program of energy

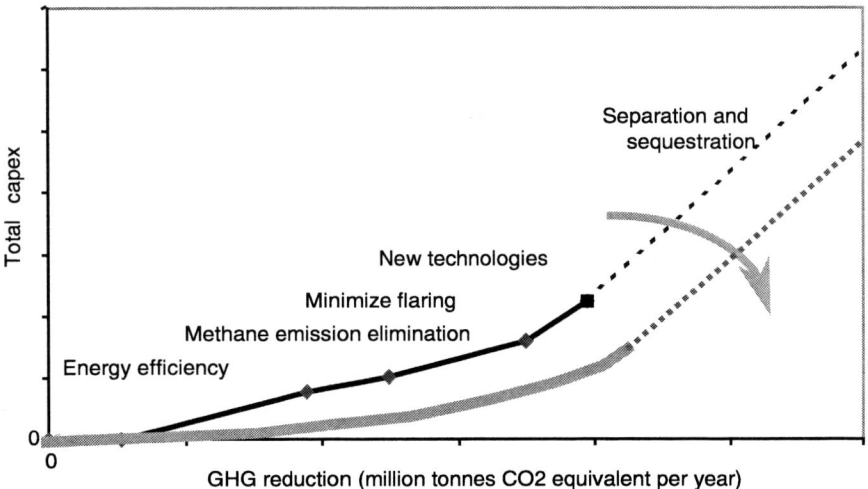

FIGURE 2.1 Relative capital expenditures needed for greenhouse gas emission reductions. *y*-axis represents relative capital investment: *upper curve*; technology of 1990; *lower curve*; technology of 2000.

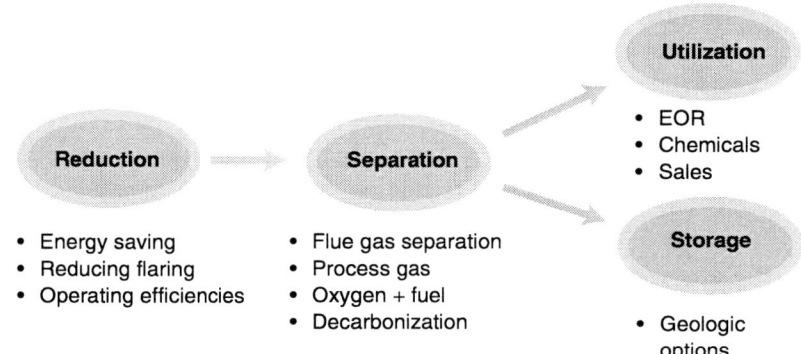

FIGURE 2.2 Strategic options for CO_2 reduction.

management. We have found numerous opportunities to minimize emissions through changes in processes, equipment, or procedures. A few examples follow:

• Site energy audits and implementation of energy management technologies have reduced refinery emissions by 5% through careful attention to energy consumption and measurement of energy performance parameters.

• One business unit saved $265,000 per year and reduced emissions by 8,000 tonnes a year by challenging the practice of maintaining spare capacity in on-site electricity generators. It developed procedures to balance the load between two turbines rather than running lower loads on three. The procedure includes regular rotation of the load and shutdown of the spare turbine. The procedure allows longer intervals between maintenance, less wear, and routine maintenance on the spare turbine while keeping it ready for rapid startup.

• Fired heater tubes can be fouled with certain crude oil components, thus reducing heat transfer efficiency, increasing pressure drop, shortening run times, and increasing operating and maintenance costs. Specially designed springs installed in the tubes increased heat transfer by 50%, reduced fouling by 70%, and doubled run length. Fluid passing through the tubes makes the springs vibrate so that they continuously scrape the inside of the tubes, keeping the tubes cleaner.

• Boiler and fired heater tubes develop scale on the fired side of the tube. Past procedures required furnace shutdown for cleaning on a regular basis. On-line cleaning with combustible abrasives allows treatment without shutdown. With on-line cleaning, one refinery experienced a CO_2 emissions reduction of 1,800 tonnes per year, $60,000 fuel per year savings, $300,000 per year yield improvement, $800,000 per year throughput increase, and an overall 1.5% improvement in efficiency for the unit.

• A petrochemical complex implemented improved divided wall column technology. Energy efficiency increased by 30%, CO_2 emissions were reduced by 30%, and capital equipment costs were reduced by 10%.

• New turbine technology that allows us to take advantage of the pressure drop as fluid is brought into a terminal is generating 3 MW(e) (megawatts of electrical power) of extra electricity from previously wasted energy.

• BP participated in the development of a new biphase turbine that converts reservoir energy to shaft power by passing reservoir fluid through a multistage turbine. It saves 10% in weight and area in cramped offshore platforms.

• Flaring (burning) of waste hydrocarbon streams has been routine practice from the earliest days of the oil and chemical industry because it is the safest disposal method. When CO_2 emission reduction targets challenged this routine practice, opportunities to reduce emissions while enhancing revenue were found. An offshore business unit aggressively reviewed its procedures and found larger volumes of salable gas being burned than previously believed. Ways to capture the gas were developed and are being implemented. The business unit is well along in eliminating all nonemergency flaring while generating substantial extra revenue. Full implementation will reduce CO_2-equivalent emissions by more than 2 million tonnes per year.

- Replacement of conventional gas-actuated valves to control flow from gas wells with no-loss systems eliminated cold venting of methane from a large production operation. This change gave the business unit an additional 5 million standard cubic feet per day of sales gas.

These examples are only a few of those found within BP. Other companies evaluating their operations through the filter of a CO_2 mitigation challenge will undoubtedly find similar improvements they can make. Energy efficiency improvements have allowed BP to reduce its emissions by more than 5% in the past two years. Efficiency improvements and process changes provide substantial reductions in emissions and play a major role in CO_2 mitigation. However, they will not be sufficient to reduce world emissions to the needed levels. This reduction will require a combination of energy efficiency, capture, and storage of large volumes of CO_2 for very long periods.

CO_2 SEPARATION

Norway's Statoil operates the only purpose-built plant designed to capture CO_2 and to store it in geologic formations. Norway implemented a carbon tax system in the early 1990s as part of its response to climate change concerns. The Norwegian carbon tax provided the economic driver to motivate Statoil and its partners to invest in a platform, separation and compression equipment, and an injection well. Carbon dioxide injection began in 1996 with approximately 2.2 million tonnes of CO_2 being stored by late 2000. The plant separates CO_2 from natural gas produced from the Heimdal formation in the Sleipner field to increase the fuel value of the sales gas to meet customer and pipeline specifications. Normal practice would have been to vent the separated CO_2 to the atmosphere.

CO_2 separation plants intended specifically for large-scale CO_2 separation and storage have not been built. Several vendors build and sell plants for relatively small applications, such as food and chemical processing. These plants have a capacity range of 100 to 1,000 tonnes per day. These processes use a basic chemical—usually an aqueous organoamine—to interact with the acidic CO_2. The resulting mixture is heated to recover the amine and CO_2 streams. Costs depend upon the required CO_2 purity and availability of process heat used in the separation process.

CO_2 mitigation plants will have to be 10 to 100 times larger than present plants. We believe that at least 1 million tonnes per year (about 2,740 tonnes a day) is the minimum capacity that will provide the needed economies of scale and that world-class plants will be larger than 4 million tonnes per year. Preliminary engineering estimates suggest that the cost of separating CO_2 from combustion gases ranges from around $65 to more than $200 per tonne of CO_2 separated when capital and operating costs are included. Definitive cost estimates are inherently site and process specific. BP believes that the cost of greenfield separation facilities will have to provide CO_2 capture for $15 to $20 per tonne to be economically viable.

BP is leading a systematic evaluation of the options for CO_2 capture from combustion processes in cooperation with a group of other energy companies.[8] The joint industry project has attracted international energy companies and government attention. The Carbon Capture Project (CCP) is a 3.5-year project to investigate CO_2 capture from combustion processes and to develop the criteria for safe storage of CO_2 by geologic means. Its objectives are to reduce the cost of capture in new construction to 25% of present technology and, in retrofit projects, to 50% of present technology costs.

Figure 2.3 is a schematic representation of the feasible combustion processes. Fuel can be burned in air to generate power and heat. Power generation methods include direct-fired heaters used to heat process fluids, boilers used to generate steam for electricity generation and process heat, and gas turbines used for electricity generation and shaft power. In normal practice, the combustion products (flue gases) are vented to the atmosphere. CO_2 capture from flue gas requires handling large volumes of hot, wet, and corrosive gases at low pressure. Flue gases range from about 2.5 to 17% CO_2 by volume, so the handling equipment is very large. Industrial operations were not designed with capture of flue gas in mind; therefore the sources are usually dispersed around the complex where they fit best to meet the design objectives.

The only commercially proven process for postcombustion separation is an amine-based process. Substantial research is needed to improve postcombustion processes and reduce their costs. Reduction in size and weight of the systems is an important objective when remote and offshore operations are considered. Improving the contact efficiency and mass transfer between the gas phase and the separation chemicals is needed. Novel methods that do not depend on large volumes of expensive chemicals would be very attractive because of industry's concerns about the waste products of the separation process.

A second combustion approach is precombustion decarbonization (PCDC) to produce hydrogen for subsequent combustion to generate heat and power. PCDC has an advantage over postcombustion because hydrogen can be produced in a central facility where the CO_2 capture process will be easier. The most common current production method is steam reforming of a high-hydrogen fuel such as methane to generate hydrogen and CO_2. The hydrogen can be burned in existing equipment with modest modifi-

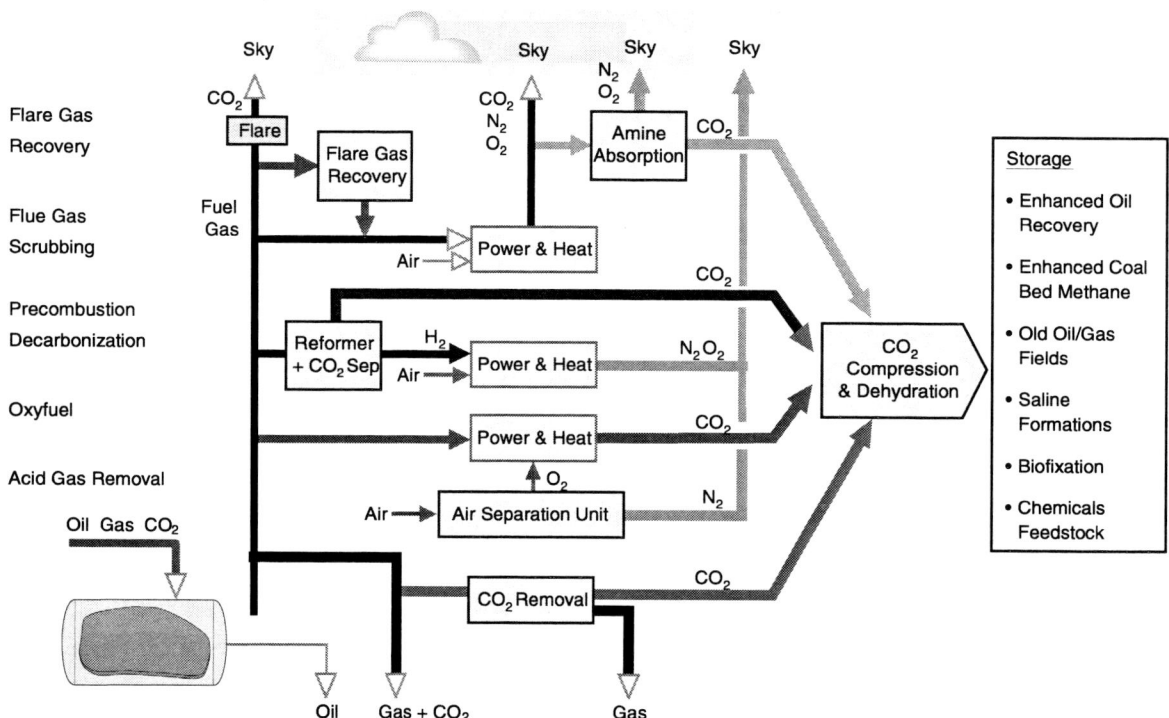

FIGURE 2.3 Separation and storage options for carbon dioxide.

cations. Research needs in hydrogen combustion include development of better, more efficient reforming processes that reduce cost and improve CO_2 capture; hydrogen-burning fired heaters, boilers, and turbines; improved materials resistant to hydrogen embrittlement and degradation; high-performance fuel cells to replace turbine-generator sets; and other novel approaches to using hydrogen efficiently.

A third combustion approach is to burn the fuel in an oxygen or oxygen-enriched atmosphere—the oxyfuel, or oxygen fuel combustion process. This method generates a high-concentration CO_2 stream that can be processed more easily for subsequent handling. Oxyfuel processes depend on precombustion separation of oxygen from air to produce the necessary high-oxygen-content stream. Research needs in oxyfuel processes include development of advanced materials to handle the higher flame temperatures of direct oxygen firing, development of advanced turbines and fired heaters with recycle capabilities to allow flue gas recycle to combustion control, methods to seal conventional furnaces to minimize mixing of unwanted air with the oxyfuel mixture, and improved methods of oxygen separation from air.

Acid gas removal from natural gas at the production facility is another major source of CO_2 emissions. Most natural gas contains some CO_2, with concentrations ranging from near zero to close to 100%. Energy companies regularly handle gases that contain up to 50% CO_2. The Sleipner field mentioned above contains about 11-12% CO_2. Pipeline specifications may require that the CO_2 content be reduced to less than 1%. Normal practice would be to separate the gases and vent the CO_2 to the atmosphere. Since produced gases are normally at high pressure, the size of the separation plant is smaller and the costs of separation are consequently lower. Processes and costs are well established, so research needs are modest. Improved column packing, solvents, and regeneration processes are areas that need improvement.

Once CO_2 is separated from other flue gas components, it must be compressed and transported to a storage site. The large volumes make transport by pipeline the lowest-cost option. Several pipelines in the United States transport large quantities of CO_2 from naturally occurring sources to enhanced oil recovery projects in the western United States. These pipelines operate at approximately 150 bar so the CO_2 is transported as a supercritical dense-phase fluid. This technology is well established and needs little technology development. Other transport methods, such as liquid by truck, rail, or ship, are possible, but the costs of such operations would radically affect the cost of CO_2 mitigation.

CO_2 STORAGE

BP and others in the energy industry are concentrating on geologic storage methods. We are most familiar with technologies needed for successful implementation because of our use of CO_2 in enhanced oil recovery (EOR) operations. We understand the processes that determine whether or not a reservoir will hold CO_2 for long periods well enough to feel confident with the method. However, there is need for research into sealing mechanisms and how they might be affected by CO_2 over long periods. Monitoring of long-term storage sites is another area in which technology development is needed. Other methods such as biofixation and chemical feedstocks hold promise but require research before they could play a significant role in CO_2 mitigation. Figure 2.3 lists a series of options for storage of large quantities of CO_2.

EOR provides an opportunity for beneficial use of CO_2 as part of a storage program. Enhanced oil recovery and enhanced coal bed methane (ECBM) have the most potential in the near to mid-term (0-50 years) because the CO_2 is used to improve production from assets that otherwise would have marginal value. The revenue stream developed from the increased or continued production can pay for the cost of CO_2 separation, transportation, and storage. Present EOR practices will have to be modified to provide long-term storage (>1,000 years). Technology convert from EOR to storage is needed and under development. Availability of CO_2 in areas too far from natural sources of CO_2 for EOR projects may lead to recovery of additional oil from well-known fields.

Coal formations offer another opportunity to store CO_2 because it adsorbs strongly on the coal surfaces. Contacting coal surfaces with CO_2 via flue gas injection may have application at coal-fired power plants near coal mines and in other areas where suitable coals are accessible. Since CO_2 adsorbs more strongly than methane on coal surfaces, there is the potential for increasing methane production from coal bed methane operations. The effectiveness of adsorption processes will depend heavily on the type and permeability of the coal formations. BP operates an ECBM project in northwestern New Mexico and southwestern Colorado. We are adding a commercial-scale demonstration of CO_2-nitrogen injection during 2001-2002.

Saline water-bearing formations are broadly distributed across the United States and other parts of the world. The Sleipner project mentioned earlier is one example. Similar projects are under consideration elsewhere in the world. Little attention has been paid to saline aquifers by the energy industry since they have no commercial value. Research efforts are needed to determine the geologic and fluid transport properties of saline water-bearing formations so that their applicability for long-term storage can be determined.

Depleted or nearly depleted oil and gas reservoirs may provide other geologic targets for long-term CO_2 storage. Inactive fields could be reopened and used for direct storage projects. A concern is that old well bores may have not been sealed to present-day standards and may have degraded casing or well hardware left in place. Considerable evaluation would be required before they could be put into service.

CHEMICAL INDUSTRY CHALLENGES

Biofixation of CO_2 includes agriculture, forestry, soil science, and enhanced photosynthesis as potential methods for carbon fixation. Photosynthesis is the most effective carbon fixation method available and is the basis for life on our planet. It plays a major role in climate change mitigation. Managing croplands, forests, and grasslands is a short-term way to capture large amounts of carbon dioxide. Unfortunately, the CO_2 can be released back into the atmosphere nearly as quickly as it was stored if management practices are changed. Deforestation is recognized as a driver in climate change.

Carbon dioxide as a chemical feedstock has been under study for many years. Most of the processes evaluated are highly endothermic and require large amounts of energy input for successful use. Substantial research would be needed to make CO_2 a significant feedstock in the chemical industry.

The recent U. S. Department of Energy publication *Carbon Sequestration—Research and Development* is the most current review of overall technology needs in carbon dioxide mitigation and possible chemical pathways to mitigation.[9] It is the result of a 2.5-year effort by a largely volunteer team of scientists and engineers active in the field. Halmann and Steinberg have published an exhaustive review and evaluation of mitigation technologies, including chemical processes.[10] These two references would provide good starting points for researchers entering the field.

REFERENCE

1. http://www.bp.com/_nav/world.htm, the section of BP's homepage dedicated to world issues for information regarding our climate change and green activities.
2. http://www.bpsolar.com for an overview of BP Solar's activities.
3. See http://www.h2interactive.net/ for information about this innovative meeting held October 11-13, 2000, in Toronto, Canada.

4. http://www.bp.com/pressoffice/bpconnect/ for a description of the latest concepts in fuel marketing and http://www.bp.com/cleanerfuels for a discussion of cleaner fuel introduction around the world.
5. http://www.bp.com/alive/index.asp?page=/alive/performance/health_safety_and_environment_ performance/issues/climate change for a discussion of emission trading issues.
6. See Chapter 1.
7. http://www.bp.com/alive/index.asp?page=/alive/performance/health_safety_and_environment_ performance/issues/climate change for a discussion of emission trading issues.
8. www.co2captureproject.org for details of the project.
9. Carbon Sequestration—Research and Development, U. S. Department of Energy, Office of Science and Office of Fossil Energy Washington, D.C., 1999.
10. Halmann, M.M., and M. Steinberg. *Greenhouse Gas—Carbon Dioxide Mitigation: Science and Technology*. Lewis Publishers, Boca Raton, Fla., 1999.

DISCUSSION

Alex Bell, University of California at Berkeley: The reutilization of CO_2, particularly by the chemical industry, is hampered by the fact that you have to add hydrogen to the CO_2 to make useful chemicals. Reformulating a carbon source to make hydrogen just adds to the problem; using water as a hydrogen source takes energy. Do you see a resolution to this problem?

David Thomas: Once the CO_2 is formed, it is very difficult to convert to a more active form without the addition of large amounts of energy. The most reasonable approach that I can think of is to use a syngas reaction to make carbon monoxide and hydrogen. The hydrogen could be used as fuel, while the carbon monoxide could be used with other reactants to generate intermediate feedstocks. I have not done thermodynamic or mass-balance calculations to determine whether this approach makes sense. My question to the chemical community is, Could such a route be feasible?

Chandrakant Panchal, Argonne National Laboratory: The 10% goal for reduced greenhouse gas emission by 2010 is an ambitious goal for a refinery. For an average refinery, it takes 450,000 Btus (British thermal units) to process one barrel of crude oil, which is equal to 5 to 9% of the energy value of the crude oil. Is the reduction 10% of your current CO_2 emission? If it is, does selling a refinery, which reduces your emission by the amount of that refinery, count toward this reduction?

David Thomas: The 10% reduction target is computed against a baseline of our equity direct emissions of all our operations. These are emissions that result from the manufacture of products that we unambiguously own and are able to affect. They do not include the total CO_2 content of all products sold in commerce.

Your second question is related to the handling of a sale or divestment of a company asset and how that affects our target. This basis is also affected by acquisitions. BP has established firm rules on the handling of these situations based on materiality. Small divestments or acquisitions (less than approximately 5% of our total) do not change the baseline. The range between 5 and 10% is handled on a case-by-case basis. Those over 10% will result in an adjustment of the baseline and the overall target. The view that we could sell several high-emitting assets to meet our goal is incorrect. Such a sale would reduce the gross emissions but would also reduce the baseline.

Acquisitions require us to determine a baseline for the acquired asset, which can take considerable effort depending upon the availability of information about the emissions of the acquired asset. Very few assets have hard data on their 1990 emissions. Obtaining valid estimates of these emissions is very complex and difficult.

In an effort to be transparent, BP has commissioned external and independent audits of its emissions by internationally known auditing firms. Their names and results are posted on our external Web site.

Tobin Marks, Northwestern University: Are there data for the chemical industry in terms of what percentage of the products go through a syngas route? Is this a major pathway?

David Thomas: I participated in several reviews of our chemical and refinery groups during which the chemistry being practiced was discussed. A fair number of them used syngas processes to generate hydrogen or to alter the chemical species. I haven't attempted to determine how many or what the percentage actually is, but I think it is a substantial number. This suggests a fertile area for research.

Dahv Kliner, Sandia National Laboratories: Could you provide further details of your internal carbon-trading experiments? What is the range of costs? What ended up being the most common way you reduced emissions? Were there any surprises in the approaches you ended up taking?

David Thomas: The trading program is in its early days. Trading was piloted with a limited number of business units in 1999. In 2000, the program was opened to all business units. During 2000, approximately 2,000,000 tonnes of CO_2 were traded with an average cost of less than $10 per tonne. Most of the CO_2 reductions being done to generate trading credits resulted from improvements in energy management or process changes. These modest changes are not too expensive and frequently generate additional product or reduced costs. Reduction of flaring by one business unit gave it an additional 5 million standard cubic feet per day of gas to sell! None of the reductions have come from new or newly implemented capture and sequestration activities. They are still immature technologies. Our biggest surprise was the lower-than-expected cost of the program.

Tom Brownscombe, Shell Chemical Company: We have been asked publicly about similar internal reductions and so on. I think we are actually ahead on the 10% reduction goal, having almost achieved it. So we are looking intensively at the same sorts of questions that you raise here.

I wanted to make one point concerning Alex Bell's comment. Opportunities to use CO_2 without having to use hydrogen are obviously of great interest. We have a set of patents coming out on the use of CO_2 to make chemical intermediates for polymers that don't involve the use of hydrogen. I think these will probably issue in the next couple of months.

Jack Solomon, Praxair: Can you give us examples of some of the energy efficiency things that have been done?

David Thomas: Several examples are shown in my vugraphs:

- Improvement of valve actuators to minimize venting of methane on a large number of producing wells
  - Consolidation of power generation in certain fields
  - Shutdown of running spares, which has saved considerable energy and reduced emissions
  - Process and procedure changes to improve operations and reduce emissions

- Improved maintenance (looking at maintenance through the lens of CO_2 emissions opened up some opportunities)
  - Improved burner efficiencies and heat exchange processes

Glenn Crosby, Washington State University: First, a comment. I certainly am impressed with trading credits. It also gets at what is easy to do, but that's finite. In other words, you'll run through this pretty quickly if, as is suggested, two-tenths of a gigatonne of carbon is consumed as the feedstock of the chemical industry annually. Do you know if this includes all chemical production or just the chemicals themselves?

David Thomas: To be precise, I would have to return to the original reference. My recollection is that it is the chemicals constituting the sum of the seven major chemical precursors produced from petroleum production. I don't recall if it included all chemical production or fertilizers.

Glenn Crosby: Could you give me some feel for how many gigatonnes of carbon are used in the production of energy relative to this—that is, just for power generation, not even the transportation?

David Thomas: The best I can say offhand is that approximately 40% of the 6 gigatonnes of anthropogenically emitted carbon annually comes from transportation, around 45% from power generation, and the remaining 15% from all other sources.

Glenn Crosby: So there is approximately 30 times as much carbon used for power and transportation than there is actually in the production of chemicals.

David Thomas: I believe so. The chemical industry is a relatively small emitter compared to transportation and power generation.

Alan Wolsky, Argonne National Laboratory: In the units that come to my mind, roughly 3 quads per year are burned under distillation columns. This is the principal component of energy consumption and concomitant CO_2 generation from the production of organic chemicals, including plastics.

3

An Industry Perspective on Carbon Management

Brian P. Flannery
ExxonMobil Corporation

Over the past two decades at ExxonMobil, my colleague Haroon Kheshgi and I have invested considerable effort toward understanding climate change. We have paid particular attention to questions of carbon sequestration and the carbon cycle. I would like to provide some information and views that put these issues fully in perspective.

I believe that in the United States, we are beginning an important debate about what the role of the private sector, federal government, and the research community ought to be in addressing the climate change issue. This meeting provides an excellent opportunity for the research community to consider its role in light of the overall issues.

In this context, it may be useful first to recall some powerful examples from the past where the government was heavily involved with energy technology. Two instances that come to mind are nuclear power and synthetic liquid fuels from coal and shale-oil. Both positive and unfortunate lessons come from these exercises. In the climate change debate, we face the specter of these negative consequences happening again. What is the role of research and development? What are the appropriate roles of academia, government labs, and the private sector?

It is becoming important to consider these questions. If climate change proves to be serious over the coming decades and requires a transition to new technologies, those technologies are not likely to be straightforward extensions of ones we know or understand today. On that point, I echo the contribution of Jae Edmonds strongly. These new technologies will have to be "megatechnologies" (i.e., integrated, interacting systems of technologies working within entire infrastructures). Furthermore, they would need to work globally, not just in the United States or in the Organization for Economic Cooperation and Development (OECD). This poses additional challenges. For example, in many ways, the existence of the Kyoto Protocol has been a fundamental stumbling block to action, since it poses such a difficult political challenge and barrier to global involvement.

I frame my discussion in this chapter on the scope of the challenge in the context of the carbon cycle and the global economy, addressing the technology and infrastructure that might be required. To finish, I provide some conclusions and issues. In particular, I address the following:

- The magnitude of these new technologies;
- Potential rates of penetration of new technology into the global economy; and
- Criteria for new technology to enter the marketplace on a large scale.

THE CARBON CYCLE

First I want to review some relevant information regarding the global carbon cycle and the processes that affect atmospheric concentrations of carbon dioxide. There are vast reservoirs of carbon in the system (see Figure 3.1) that can exchange fairly rapidly with the atmosphere, which contains about 750 gigatons (1 gigaton = 10^9 tons) of carbon (GtC). The terrestrial biosphere and soils contain about 2,000 GtC; the mixed layer of the ocean contains about 1,000 GtC; and the deep oceans, 38,000 GtC.

These numbers—averages for the 1980s estimated by the Intergovernmental Panel on Climate Change (IPCC, 1995)—are reasonably well determined, especially in the context of this subject. (There are also larger reservoirs in which exchanges occur on geological time scales, but I do not discuss those.)

Exchange fluxes of carbon among these systems must also be considered. The human contribution to emissions to the atmosphere from combustion of fossil fuels—about 5.5 GtC per year in the 1980s and about 6.3 GtC per year in the 1990s—is reasonably well known to within about ±10%. However, estimates of net emissions from tropical deforestation, shown here as 1.6 GtC per year, are far less reliable. Again, the values I am citing are from IPCC (1995), in the case for the decade of the 1980s. Figures for land use change as a whole, especially considering reforestation in the middle latitudes, are less certain. Nonetheless, the consequences of tropical deforestation are important.

In addition, there are vast natural cycles involving two-way exchanges of CO_2 into and out of the terrestrial biosphere through respiration, photosynthesis, and decay. Carbon dioxide is also exchanged into and out of the ocean through the mixed layer, by thermodynamic processes of gaseous invasion and

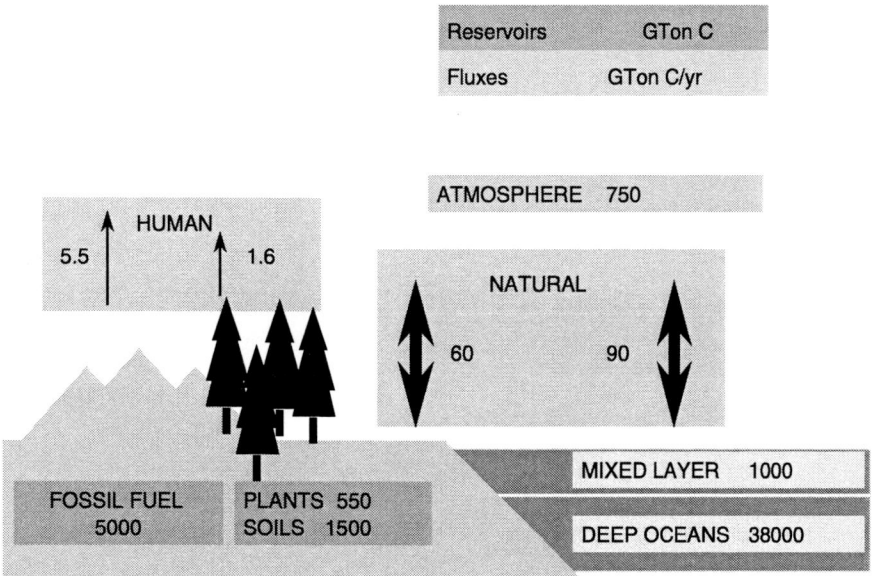

FIGURE 3.1 The carbon cycle. SOURCE: IPPC (1995).

evasion. These processes amount to about 60 and 90 GtC per year, respectively, but these numbers are probably accurate only to within 30-50%.

Although we know the human emissions fairly well, we don't know the natural emissions well at all. Added to this uncertainty is the fact that natural emissions can change as a result of long-term climate changes. From data on the year-to-year fluctuations in the accumulation of atmospheric CO_2, it appears that they can also change as a result of volcanic eruptions, fluctuations in sunlight, and other factors this may not be understood. These factors make understanding CO_2 in the atmosphere difficult. Adding to this difficulty is what might happen in the atmosphere over the next 100 years if these processes themselves begin to change.

Figure 3.2 shows estimates of human emissions of CO_2 from use of fossil fuels in 1990 and from estimates of emissions taken from the IPCC IS92a scenario (IPCC, 1992). This scenario is described more completely in Chapter 1. For our purposes, over this relatively near-term period, the results differ only slightly from scenarios produced by a variety of forecasting agencies. They are similar to projections of what ExxonMobil might produce for the next 20 years. Over this relatively short period, projections depend on trends and technologies that are reasonably well understood. However, for discussions of climate change, the most important times are after 2020, through the next century, and beyond. Here, we all should be very humble about trying to make projections.

Two powerful conclusions can be drawn from the scenario. First, emissions are going to grow rapidly to meet the demands of society for prosperity and to meet basic needs. Critical assumptions that enter these are population growth, which is discussed in earlier chapters, future rates of economic development, and technology change.

The daunting challenge is that emissions are growing most rapidly in developing countries. Countries such as India, China, and Indonesia are going to rely on domestic coal to meet growing needs, especially for electric power, and their emissions are going to grow rapidly. Over the next 30 years, the installed technologies are going to be based on what we know about or can foresee. Over the next 100 years, this may or may not be true, given the different kinds of changes that can occur. Next, let us focus

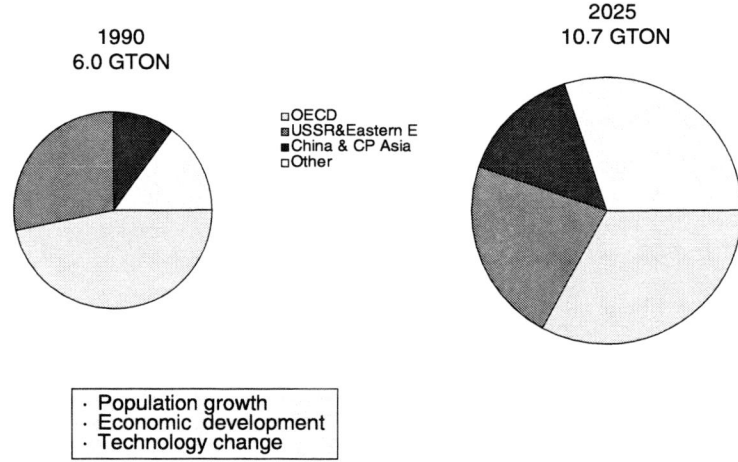

FIGURE 3.2 Regional fossil fuel CO_2 emissions. SOURCE: IPPC (1992).

on recent projections for CO_2 emissions in the United States from fossil fuel use without including sinks from land use or emissions of other greenhouse gases

Figure 3.3 is based on information from the U. S. Energy Information Administration (EIA, 2000a,b). In 1990, emissions were about 1.4 GtC per year. The black squares show actual annual emissions. In 2000, emissions in the United States will already be 24% above the Kyoto target, which begins only eight years from now. The insert shows emissions in 1997 broken into three classes—electric power use, transportation, and all other uses combined. In total, these three add up to U.S. CO_2 emissions in 1997. Reductions in any one of these classes alone, even if emissions were eliminated completely, would still not allow the United States to meet its target in the year 2000. That's the scale of the challenge. How quickly can you change the infrastructure? How do you achieve this? Note, in the figure, that there was one downturn in annual emissions—1991, a year in which the economy was in recession. Even the recession produced only a slight downturn.

Next, I want to place the consequences of the emissions reduction called for in the Kyoto Protocol in the context of the full climate change problem.

Figure 3.4, based on simple conventional models shows the possible effects of Kyoto from 1990 through 2100. The results are based on the IS92a scenario, but introducing the Kyoto limitations. In this case, I assume that the developed world reduces its emissions to 5% below the 1990 levels and holds them at this point for the next 100 years and the developing world continues to emit as projected in the IS92a scenario. These are crude approximations. The top curve shows how temperatures might evolve under IS92a. The lower curve shows the consequences of Kyoto. The net effect is to delay the projected temperature rise in 2100 by approximately a decade. The chart should not be taken too literally because the uncertainty associated with the temperature rise is larger than the scale in this graph.

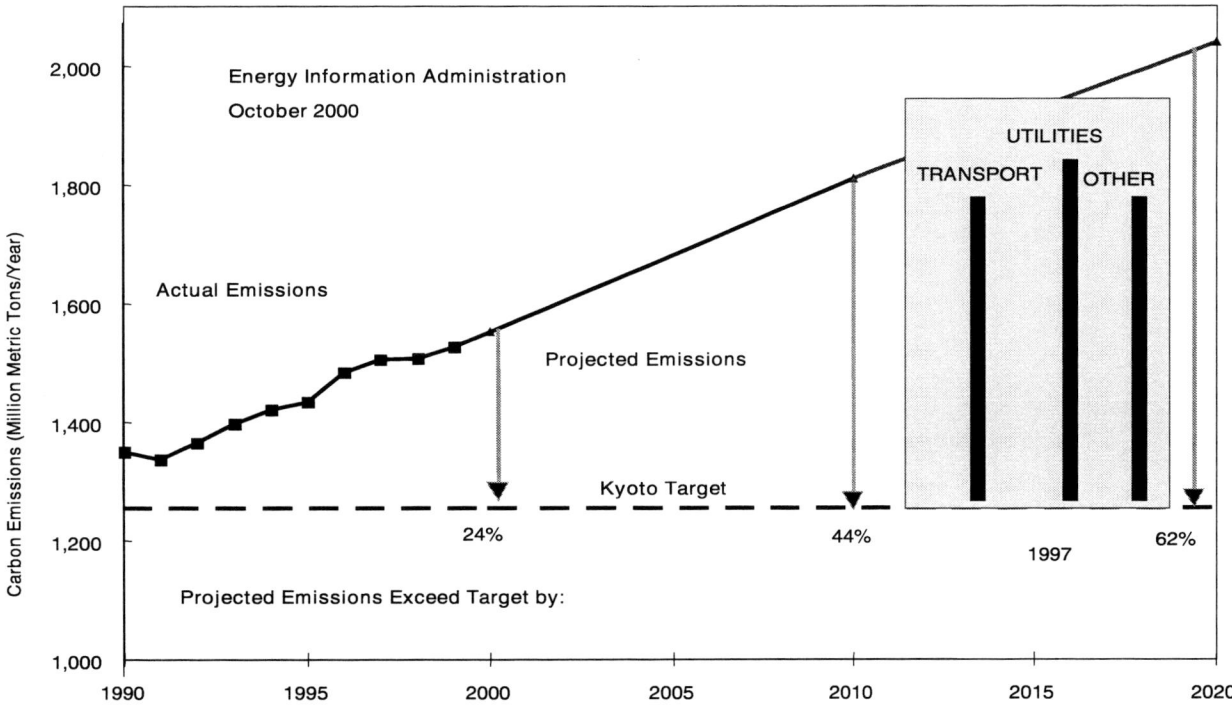

FIGURE 3.3 U.S. carbon emissions: projected versus the Kyoto target. SOURCE: EIA (2000a,b).

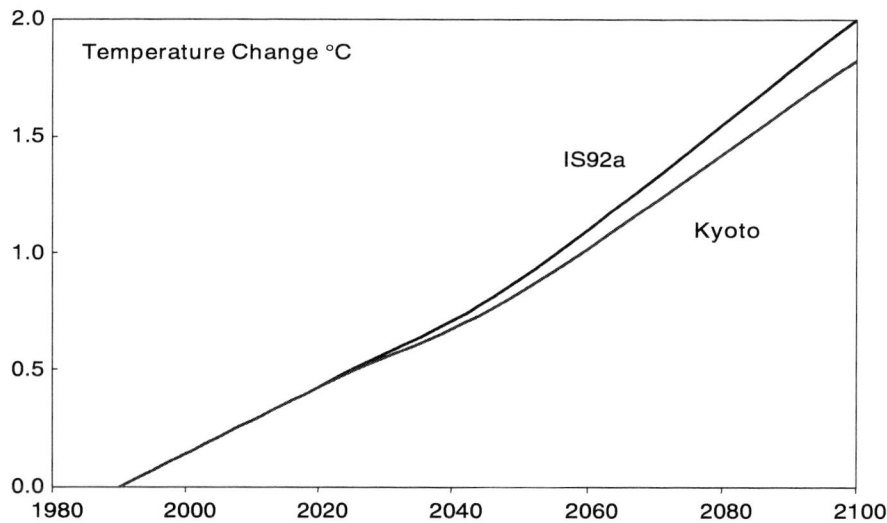

FIGURE 3.4 Climate implications of Kyoto.

Although society would probably be unable to detect the climatological differences between the two projections, the differing economic consequences would be evident. While it is not my point to say that the Kyoto Protocol has no effect, in terms of addressing the climate change issue, Kyoto does very little to address global CO_2 emissions. The conclusion is that more difficult steps would be required if climate change proves to be serious and that these steps must address emissions in developing, as well as developed, countries.

My final information from the climate change arena is shown in Figure 3.5. It further elaborates what the response might mean if society determined that it must stabilize atmospheric CO_2 concentrations. Again, the information presented here complements material discussed in Chapter 1. The figure shows what track emissions would have to take for stabilization to occur at a given level. It is divided into two families of curves. The lower family of curves shows emissions only from the so-called Annex 1 countries—developed countries that agreed to emissions commitments in the Kyoto Protocol negotiations. The top family shows three projections of global emissions. In the upper group, the highest curve, rising to about 14 GtC per year, results from the now-familiar IS92a scenario through the year 2050.

The middle curve results in atmospheric stabilization at 550 parts per million (ppm) CO_2. The lowest curve in the top family ultimately results in stabilization at 450 ppm CO_2 globally. Jae Edmonds has described other scenarios. For example, you could devise other scenarios for stabilization at 550 and 450 ppm, in which emissions remain higher in early years but would have to fall more rapidly in later years to stabilize atmospheric CO_2 at a certain concentration.

The topmost curve in the bottom family of emissions corresponds to Annex 1 parties' emissions under IS92a. The next two curves in this family represent what emissions from Annex 1 countries would have to be if developing countries accepted no emissions commitments and the world's target CO_2 concentration was 550 ppm or less.

In ongoing negotiations under the United Nations Framework Convention on Climate Change (FCCC), the European Union has taken the position that the Kyoto Protocol should control CO_2 emissions such that concentrations will stabilize at 550 ppm or less. The target of 550 ppm was formulated politically. As its objective, the FCCC calls for stabilization of greenhouse gas concentrations at a level

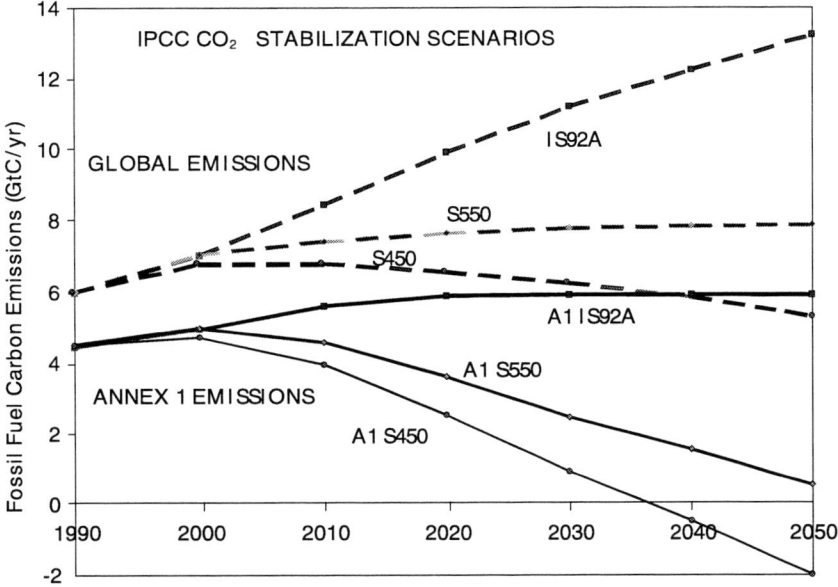

FIGURE 3.5 Emission levels required to reach the European Union proposal to stabilize cumulative atmospheric CO_2 concentrations at 550 ppm or less.

that prevents dangerous human interference with the climate system. However, science today cannot draw conclusions about the level of greenhouse gases that would be appropriate as a stabilization target.

The situation is further complicated by the fact that gases other than CO_2 are greenhouse gases. If we are worried about climate change consequences at the equivalent of 550 ppm CO_2, we must take into account methane and other gases. It may be necessary to keep CO_2 below 550 ppm to keep to the climate change consequences associated with that level of CO_2.

What would be the consequences for stabilization of CO_2 emissions at 550 ppm or less? To achieve this level without obligations by developing countries would require a phase-out of the use of fossil fuels by the middle of the century in the Annex 1 countries. The scale of the problem is enormous. It would require large-scale development and deployment of new technologies that are currently noncommercial. It is clear that at some point, participation by developing countries would be necessary. Achieving the commitments made by developing countries would almost certainly require the transfer of substantial resources from the developed world.

INTRODUCTION OF NEW TECHNOLOGY

To place changes in the decarbonization of global energy consumption into historical context, we can examine shifts in primary fuels that have occurred in the last 200 years. Figure 3.6 shows trends in the hydrogen-to-carbon ratio of energy use over the past 200 years.

Transitions between wood, coal, oil, and methane as fuels, represented by their characteristic hydrogen-to-carbon (H/C) ratios, required about 50 years each. The essential drivers in these transitions were performance demands, especially in energy intensity and end use. Environmental drivers were also of some importance. Although they are especially significant now, even in earlier periods environmental drivers such as maintaining a supply of readily available wood were present. The associated efficiency gains were made possible primarily through changes in the form of fuels, as well as changes in combus-

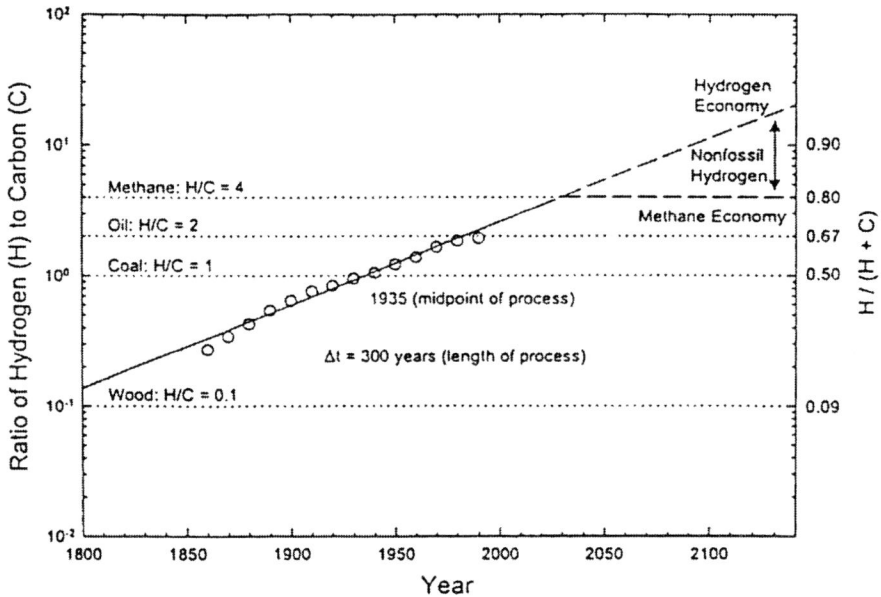

FIGURE 3.6 Two hundred year trend: decarbonization of global energy consumption. SOURCE: Adapted from Marchetti (1985); Ausubel (1996); Ausubel et al. (1998).

tion temperatures and materials. The transitions required numerous, large-scale upgrades in energy supply and end-use technology, with significant time intervals needed for these changes. Typically, more than one energy system was available at the same time. In many cases, the transition between two energy systems occurred because of obvious economic and performance advantages of the newer systems.

Figure 3.7, shows the progression in efficiency of power generation from motors. Typical doubling times are about 50 years. Materials science and materials technology have been among the key enablers in the progression. Enhanced performance for the end user made the new technology commercially attractive.

Figure 3.8 shows the growth of transportation infrastructure that enabled widespread use of the new technology for supplying and using energy. The characteristic time scale for the penetration of new technology is 50 years. In many cases, newer technologies were becoming increasingly prevalent even as old ones were still growing.

While advances in R&D enabled new forms of technology, the widespread use of a technology occurred because it provided qualitatively new service and higher levels of performance. Powerful economic drivers came about naturally through the operation of markets. The same cannot be said about many technologies being championed as potential contributors to address climate change. Rather, to maintain today's performance, new technologies could provide only the same or even reduced service at higher cost. This poses a challenge from the perspective of consumer acceptance and economic consequences.

Figure 3.9 illustrates the scale of today's infrastructure for transportation fuels and the petroleum industry. Overall capacity for today's petroleum and transportation fuels industry supports the production and refining of more than 3 billion barrels per day. This produces 1.8 billion gallons of fuel, including about 1 billion gallons of gasoline and roughly 1 billion gallons of diesel fuel. This massive

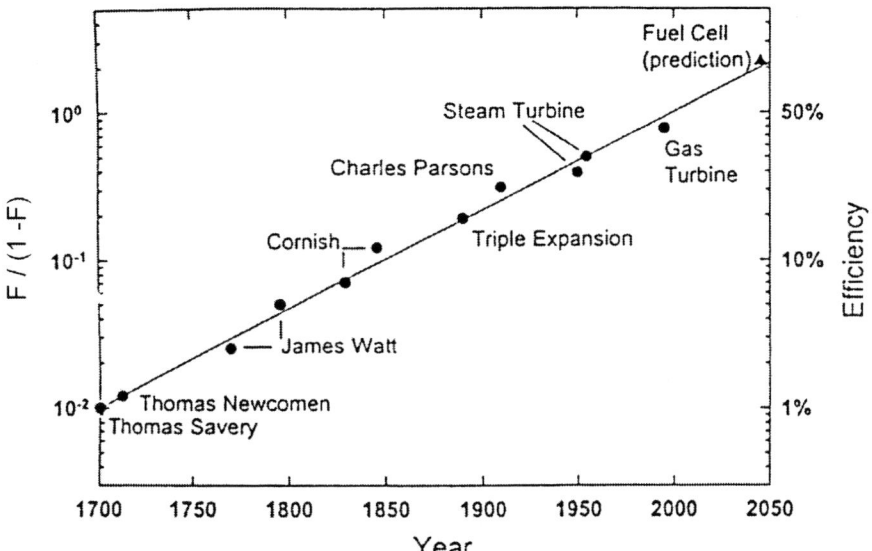

FIGURE 3.7 Efficiency of "motors": historic inventions. NOTE: F = efficiency, where 1 = 100%. SOURCE: Adapted from Marchetti (1985); and Ausubel (1996).

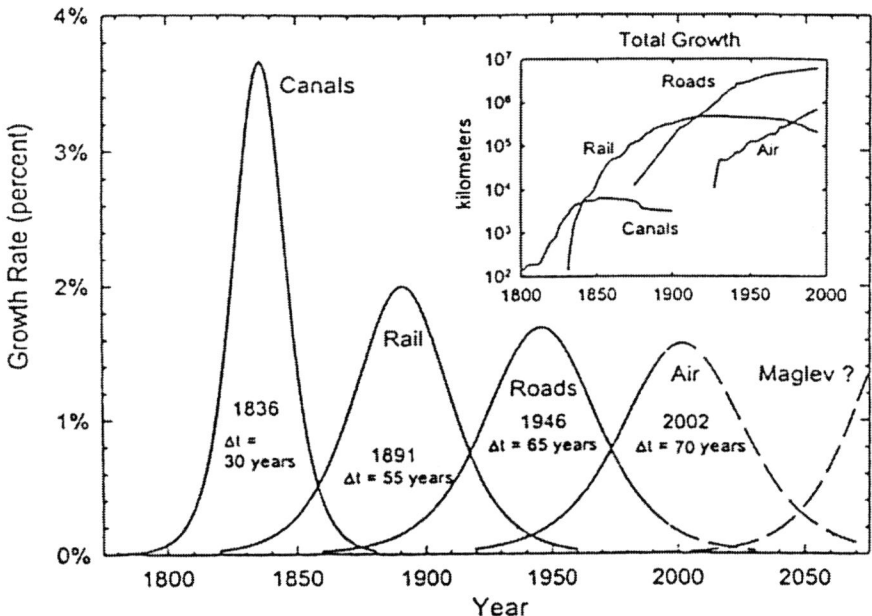

FIGURE 3.8 Penetration of major U.S. transportation infrastructures. SOURCE: Adapted from Marchetti (1985); Ausubel (1996).

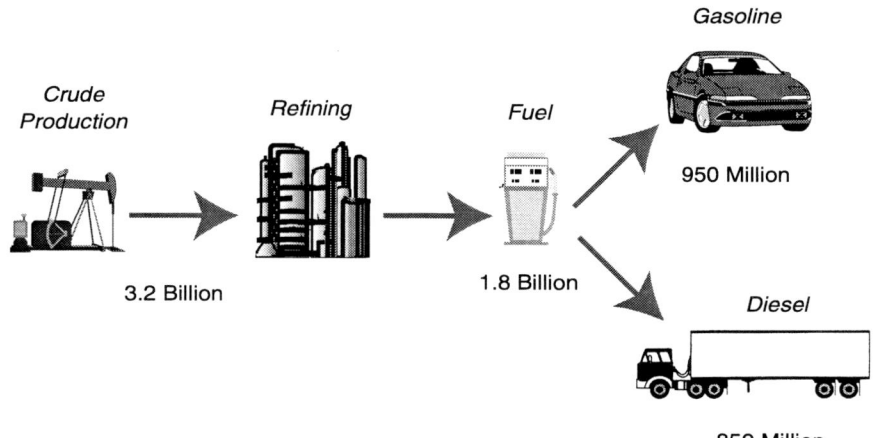

FIGURE 3.9 Today's fuel infrastructure.

fuel infrastructure was developed over a period of 100 years, including ongoing development and efficiency improvements.

From the perspective of the climate change issue, it is important to note that CO_2 emissions from the use of oil in the world's economy come primarily from end users. Of the total amount of CO_2 emitted, about 13% results from energy consumed in production, refining, manufacturing, and distribution of fuels, while about 87% comes from the end use by consumers. The petroleum industry expends considerable effort to control emissions in production, refining, marketing, and other areas because it is in the interest of the industry to do so. However, the real focus for climate change must be on the end use of fuels, not on operational emissions, if climate change proves to be a serious issue.

The discussion in this section highlights the time scales associated with widespread use of new energy technology for supply and end use. Introduction of new technology requires advances in research as a prerequisite, but widespread use requires—above all—consumer acceptance based on economic advantage. In addition, it requires investment and the introduction of essential infrastructure to support the new technology.

TECHNOLOGY OPTIONS TO ADDRESS LONG-TERM CLIMATE CHANGE

Although no one new technology will solve the entire problem, there are a number of promising options for megatechnologies that could make a substantial contribution to limiting or eliminating future emissions of CO_2. All of them have to address the challenges of economics, performance, and associated environmental impacts. Consequently, all new technology solutions require extensive research and development to address significant current barriers to their widespread commercial use.

Carbon storage in forests and soils does have the potential to make a substantial difference, but it will not solve the problem by itself. Judging from the recent failure of international negotiations at the Hague (November 2000), the extent to which efforts to store carbon in forests and soils might qualify for credits under proposed international regimes is also unclear. Nonetheless, carbon storage does offer significant potential to allow advances in the removal of CO_2 from the atmosphere.

Major technical potential exists for intentional separation and sequestration of CO_2 from large combustion facilities. In today's global economy, emissions from large facilities account for about 30% of CO_2 emissions. Electric utilities account for the largest portion of this 30%, but refineries, chemical plants, smelters, and other energy-intense industrial operations also contribute.

Analyses of the separation-sequestration approach indicate that separation is the key cost component. Among the critical design considerations is whether to combust in air or in oxygen. In either case, procedures must be designed to remove O_2 from air or CO_2 from flue gas or both. Additional procedures are needed to compress the CO_2 to high pressure in order to move it elsewhere and dispose of it for long periods of time.

In many ways, disposal is the more challenging societal question, although several options may prove to be worthy of consideration. The best options for storage of substantial amounts of CO_2 involve oceans and deep saline aquifers. While it is also possible to store CO_2 in depleted oil or gas wells, and in some cases to achieve an economic benefit through enhanced oil recovery, their capacity appears to be much smaller than that of oceans or deep saline aquifers. In addition to having a vast carbon-storing capacity, deep saline aquifers appear to be fairly ubiquitous.

Overall, economic costs are dominated by capture and transport to disposal sites. There are serious questions surrounding public acceptance of CO_2 capture and disposal. While ocean disposal may be promising, many environmental groups have already begun to mount campaigns to challenge this option. In every case, scientists and their research can contribute to the analysis of potential options for disposal and to the debate surrounding their acceptability to the public.

The capture and disposal of CO_2 from large facilities has the potential to apply to the roughly 30% of global CO_2 emissions produced by electric power plants and other large industrial facilities, but this potential can be realized only if the technology is used in both developed and developing countries.

An additional 30% of total global CO_2 emissions results from the transportation sector. Advanced vehicles hold a good deal of promise for reducing emissions. Vehicles powered by hydrogen fuel cells garner most of the attention in the context of climate change. Some version of this type of vehicle may enter the economy to a large extent in the next several decades. However, advances in the internal combustion engine and in the diesel fuel engine are also going to be significant over that time frame.

To provide fuel for fuel cell engines, several options exist, including the use of onboard reformers to convert liquid hydrocarbons to hydrogen. The essential advantage of fuel reforming would be to eliminate of the need for a massive and costly infrastructure for hydrogen production and delivery—a new infrastructure that would coexist with conventional gasoline and diesel for several decades. Hydrocarbon-powered fuel cell vehicles could hasten the widespread commercial use of fuel cell vehicles by decades compared with options that rely directly on hydrogen.

Production of low-carbon fuels from fossil fuels is receiving a lot of attention and will be addressed more fully in other chapters of this report. As an example, it is possible to generate hydrogen from fossil fuels in large facilities where CO_2 separation and disposal approaches could then be used. Although such systems present a wide array of challenges, they also present opportunities for potential synergies within the global economy. If hydrogen can be made abundantly and the means for its transport and use in end-use devices can be worked out, a synergistic relationship is created. This situation has the ultimate potential to address 100% of the power needs of the global economy at some point in the future. Such systems are not yet economical, but with research and development, it is conceivable that they could become so by the middle of the twenty-first century.

Another option worthy of research is geoengineering. In particular, the technology that shows the most promise right now is the enhancement of marine fertilization to promote removal of CO_2 from the atmosphere. I am not advocating the introduction of marine fertilization today, but I would advise that it be studied further as an option. If the consequences of climate change require dramatic action, all options would have to be considered. Geoengineering, with emphasis on marine fertilization, continues to be an option that shows some promise.

To this point, renewable sources of energy have not been stressed on this list of "mega-options." With the exception of hydropower, renewables succeed only in niche applications in today's markets. Renewable energy simply is not competitive with conventional fossil fuels in cost and reliability. Many sources of renewable energy face their own environmental problems as well. For example, using biomass and wind for energy would present environmental problems if deployed on a scale that would make a significant difference to the global economy. In estimating the cost of renewable energy it is essential to recognize that the intermittent supply produced by renewable power sources necessitates additional cost to provide conventional backup power. Such costs must be included in any realistic analysis of renewable energy.

Any of these new megatechnologies requires the introduction of significant enabling infrastructure if the technology is to gain widespread commercial acceptance. Why do large-scale infrastructure changes take so long? In addition to the technology itself, all related enabling technologies must be developed as well. Capital also must be invested to achieve substantial market penetration. Historically, in the examples of technology transition discussed earlier, public concerns were not a major factor. In those cases, the public generally encouraged the development of new technologies.

The case may not be the same today. In addition to the technical, scientific, and environmental challenges, the development of new technologies would each have a set of public perception issues to address. Consider, for example, gas handling. In a CO_2 sequestration era, CO_2 pipelines and accompanying permits would present problems because of the "not-in-my-back-yard" attitude. Hydrogen production, supply, and storage face safety concerns. The development of a given technology would face resistance from groups who advocate their own preferred approaches.

Moving Technology from Introduction to Widespread Use

The rate of market penetration of a new technology will be a function of technology development as well as the success of potentially massive investment—an uncertain prospect. The central issue may be how to introduce new technologies into a market on a small scale and then get them to grow into widespread commercial use. Most analyses envision a desirable final state, with everything up and running. A more relevant question may be, How does a new technology evolve successfully from introduction to widespread use?

The scope of the capital investment required for the transition to a currently non-existent fuel technology is truly staggering. Capital cost for fuels infrastructure alone will approach or exceed $1 trillion. For comparison, this total is 13 times the total capital employed by ExxonMobil Corporation—the largest corporation in the petrochemical industry. It is equivalent to the total annual capital investment in the U.S. economy, or to the expected real growth in capital investment over the next 30-40 years. Other priorities of society, besides climate change, require capital investment as well.

Finally, let me discuss a number of dimensions in which new technology must succeed, if it is to come into widespread global use. To be successful in the marketplace, technology has to succeed in a number of critical dimensions. Failing in any one of these areas will doom the technology in its attempt to achieve widespread commercial use. For example, battery-powered vehicles work and reduce local emissions, but they have failed in the marketplace because they don't have the performance, range or cost that people want. Even with government mandates, they have not come into widespread use, only niche use. The list of criteria that are crucial to widening the use of a technology includes the following:

- Performance
- Cost

- Safety
- Regulatory compliance
- Enabling infrastructure
- Reduced negative environmental impacts
- Consumer acceptance

For a technology to be considered, it must first be capable of delivering expected performance. A major barrier exists if newer technologies do not perform as well as existing technologies. Improved environmental characteristics are not enough to sell people on a new technology that cannot get the job done in the way that they are accustomed.

Those involved with commercial enterprise understand that competitive costs matter for the success of a new technology. Governments can subsidize small-scale costs and early market penetration, but governments are not able to bring about global use of new energy technologies. Taxes on a new technology can become a contentious issue for those that get past the initial stage. Entrepreneurs looking to develop a market for new technologies will often depend on subsidies, but as their market penetration grows, the government may remove the subsidies and subject the technology to a risk of collapse before it can make a profit on its own. Indeed, governments are more likely to tax energy products, if they come into widespread use.

Making a new technology safe for consumers, both in reality and in consumers' perceptions, is a key issue. Addressing the safety question will be especially relevant for those introducing hydrogen-powered technology to the economy, since these concerns are already well known.

Any new technology will face the challenge of complying with regulations associated with manufacture and production for market, as well as with the use of the technology and its disposal.

Enabling infrastructure for a new technology refers not only to capital infrastructure, such as roads, bridges, and pipelines, but also to personnel needed to support the use of the new technology. Education and training are necessary to prepare people for tasks involved in designing and managing the systems, performing repairs, and supplying spare parts.

Concerns about environmental impacts, land use, and access will arise when an energy technology is deployed on a global scale. For example, the use of biomass, especially as a source for liquid fuels, faces an enormous set of environmental issues.

The important point is that if a new energy technology fails in any of these critical dimensions, widespread commercialization will not occur. The research community should work to address each criterion for market acceptance when considering new technology options. While performance, cost, safety, and public acceptance are most important, researchers should aim to identify fundamental barriers to widespread use in any dimension. Research must be aware of the dimensions of performance that may become important and avoid focusing solely on emissions, separations, or power plants.

Identifying barriers, seeking solutions, and seeking public acceptance are key things that the academic and publicly supported research communities should address if these options are to work. Well-resourced constituencies exist who will actively work to create obstacles to public acceptance on economic or environmental grounds. I think this will be interesting to watch.

One significant obstacle to the introduction of a new technology is the fact that a new technology must compete against steady improvement in base technologies. For example, hydrogen-powered fuel cell vehicles must compete not with the internal combustion engine of today, but with the improved version that will be in place 20 or 30 years in the future.

If these potential new energy technologies for reducing CO_2 emissions require taxpayer support, then they must compete for research and development resources with other promising approaches to addressing global climate change and, they must compete with other research priorities. R&D options might include end-use efficiency to curb CO_2 technologies aimed at reducing greenhouse gases other than CO_2, and technologies for adaptation. These areas will be competing for public- and private-sector funding.

ISSUES, CHALLENGES, CONCLUSIONS

I would like to finish with personal conclusions as someone with a private-sector perspective. The goal of R&D in carbon management should be to create economically justified options for future technologies that will make a difference to the global energy situation. Taxpayer-funded research and development should seek to identify fundamental barriers to technology, as well as finding solutions that improve performance, cost, safety, environmental acceptability, and consumer acceptability.

Taxpayer-funded resources today should not be wasted on optimizing currently uneconomic technologies. These technologies will not enter the market substantially for many years, if they do at all. Spending tax dollars on expensive pilot and demonstration studies to optimize technologies that are not viable economically is extremely expensive and unlikely to deliver a product of lasting value.

I think this is the fundamental question concerning the role of research and development for society. Taxpayer-sponsored initiatives create opportunities for inertia, boondoggles, pork-barrel funding, and white elephants that could become a problem. This will be a challenge because it creates opportunities for big budgets and employment gains in some areas through politically motivated demonstrations of action that are unlikely to lead anywhere.

At the end of the day, the private sector should bear the risk and capture the rewards of developing commercial technology that will ultimately compete in the market. Historically, governments tend to be ineffective at supplying markets efficiently. The private sector is far more successful. Even more importantly, it is the private sector and not the government that should suffer the loss if mistakes are made.

Reducing CO_2 emissions is a vast and challenging task that, even with concerted action, can occur only over decades. To effect real change, solutions have to apply globally, and they must be carried out as science and technology continue to evolve.

It is a pleasure for me to acknowledge the contributions to this paper of my ExxonMobil colleagues Haroon Kheshgi, James Katzer, and Roger Cohen.

REFERENCES

Ausubel, J.H. Can technology spare the Earth. American Scientist 84, 166-178, 1996.
Ausubel, J.H., C. Marchetti, and P.S. Meyer. Toward green mobility: the evolution of transport. European Review 6, 137-156, 1998.
EIA. Annual Energy Outlook 2001. Energy Information Administration, U. S. Department of Energy, Washington, D.C.: DOE/EIA-0383(2001), 2000a.
EIA. Emissions of Greenhouse Gases in the United States 1999. Energy Information Administration, U. S. Department of Energy, Washington, D.C.: DOE/EIA-0573(99), 2000b.
IPCC. Climate Change 1992: The Supplementary Report to the IPCC Scientific Assessment. Cambridge University Press, New York, 1992.
IPCC. Climate Change 1994: Radiative Forcing of Climate Change and an Evaluation of the IPCC IS92 Emission Scenarios. Cambridge University Press, Cambridge, U.K. , 1995.
Marchetti, C. Nuclear plants and nuclear niches. Nuclear Science and Engineering 90, 521, 1985.

DISCUSSION

David Keith, Carnegie Mellon University: I would like you to comment on what I see as a real problem—the decline of large-scale research and development in a corporate setting. We have seen a real evaporation of the old corporate research laboratory across many industries, from electronic technologies to energy. Lucent has been effective at mining some of the core, very-long-term research that Bell Labs used to do, and as you know, investment in long-term research and development for the electric power industry has been very much attenuated. Research and development in industrial settings seems to be a vital niche between very long-term fundamental science—that the federal government can fund—and the short-term product development. Can you tell us how you or Exxon see ways to fill that gap?

Brian Flannery: It's a very good question. I don't have an easy answer. One thing I can say, and I come out of that ExxonMobil research and development community, is that at ExxonMobil, we have still held on to ours. It's smaller, but we still have research. The challenge to us is to make sure we find some way of delivering, because if we don't, we'll be gone too.

I think we can deliver. I believe that technology, the pace of technological change, and the opportunity to knit it together to create commercial value are probably as big as they have ever been, or bigger, but the question is how.

I also note that the decline of large-scale corporate research and development is changing the competitive landscape in ways that are going to make this debate we're entering into about the role of the federal government, even more difficult. Many of our competitors have abandoned their own research and development and, therefore, are looking for government partnerships and taxpayer-funded resources to accomplish what might be viewed as private outcomes. That's going to pose a challenge to everyone.

It's going to be a challenge to us, because we are hoping to deliver, through our own research and development, a competitive advantage. Yet if we are competing with people who are working in partnerships with government, it's going to be an interesting question of how all this plays out. Your question is very important. I don't know the answer, but it's one that is of interest to the entire research community.

Geraldine Cox, EUROTECH: There appears to be an interesting disconnect between public perception and public action. Europe is the most concerned about global climate effects, yet it has allocated less money for research than any other region. In the United States we say that we're concerned about global impacts, yet we are still buying larger cars that consume more gas and generate more CO_2—the biggest sellers are the sports utility vehicles, vans, and trucks. So as a society, we say one thing, and we act against our stated beliefs.

In a way, you were talking about the same thing. If we look at truly active greenhouse gases, this conference should focus on a wider field than CO_2 management. Carbon dioxide is one of the weakest climate-active gases relative to methane and some of the other gases. Yet it is more politically acceptable to focus on CO_2 than on some of the other gases, because the latter are harder to control. Are we really solving the problem by misdirecting our focus?

I believe we should approach the control of climate-active gases in the most scientifically expedient manner—not the most politically expedient manner. We must focus on the problems that can have the largest impact on the solution.

Also, the politically expedient big-industry focus should go hand-in-hand with a concerted effort to tackle the entire problem—population growth, efficient transportation, life-style changes, and appropriate industrial control on a worldwide basis.

Brian Flannery: You have wrapped up a whole bunch of things. Yes, the concern about climate change has so far been expressed in political commitments to achieve outcomes without talking about how to implement them. Credit has been claimed for making the commitment, but now the question is, What are we going to do? Even emissions trading doesn't actually do anything. It's just a means for identifying options and financing them more cheaply. At the end of the day, you've got to do real things in energy supply and conservation, such as fuel switching and new technologies, if you are going to affect outcomes. That's the challenge.

I understand that there is the belief on the part of many that if government sets clear policy goals, industry can deliver. I'm not sure that's true. Some of the mandates in the United States in transportation haven't worked, and the government relaxed the mandate after a lot of resources were wasted.

On the question of other greenhouse gases and different approaches, yes, that's a very valid point. Many of them actually are very amenable to reductions. After you get beyond CO_2, methane, and nitrous oxide, there are ways to deal with those other gases. They are powerful greenhouse gases, but they are only in small use so far.

However, although they are such attractive targets, if you look at a regulatory regime, perfluorocarbons and the like would be gone from commercial practice or contained by the end of the first commitment period, and they would not provide an ongoing means of improvement. Although this could be a very attractive near-term step, the point is you won't gain those cost efficiency improvements in the second commitment period, because the other gases will be gone. Methane offers very attractive opportunities; alternate uses for it exist. I think the scientific community is coming to the conclusion that the weights used for methane in the Kyoto Protocol may have been too low, but that was a political decision. The weight used for methane is the 100-year global warming potential. If it were the 20-year global warming potential, methane would be much higher valued in the near term, but the political process wanted to address energy use, so it used this weight intentionally. There are arguments you could make for using the 100-year global warming potential in terms of the long-scale issue, but I think if you talked about policy or technology or near-term ability to affect regulatory enforcement, you would give much more attention to methane than the Kyoto Protocol gives it.

At the end of the day, I think I would agree with what Jae and others have said after you have gotten through the other gases, if climate change proves to be serious, you will have to address CO_2 produced by using energy. The question is when, and what's the most attractive approach in the near term.

Chandrakant Panchal, Argonne National Laboratory: I think the last two questions really touch the basis of what we are talking about here. I just remember the last couple of things into which the U.S. government has put money. The Carter administration funded solar programs. We tried to develop solar technology in a very short time. There were lots of demonstration projects, but we did not get the best results from these technologies. If we don't act proactively in providing a roadmap, we will again spend a lot of money and try to do something in a short period of time, then repeat what happened before. Instead, we should be proactive. Let's do one, two, three, and then make the change accordingly, rather than keep talking about these things. Do you have some thoughts on that—how the industry can take the lead rather than waiting for direction or policy from the government?

Brian Flannery: Well, I have a couple of comments. First, I would recall the case of synthetic fuels from coal. That was an approach where taxpayers and heritage Exxon lost a great deal of money. I hope we've learned from those experiences. We don't want to see that happen again. It isn't just a question of money; it's people and colleagues and investments in whole communities that are disrupted by bad decisions when public policy leads the way and remains long after the need is gone. I think the fundamental difficulty is how fast you do something.

For the purposes of this discussion, I didn't mention the other greenhouse gases, because this workshop is about carbon management. Otherwise, we might have talked about methane a little bit more. For this meeting, I wanted to raise some broad, general questions about how research needs to be focused on widespread, possibly global, technologies for 20, 30, or 50 years from now. In my view, it's up to the research community—and yes we have several points of view, but I think everybody does—and we need to sort through them and come up with priorities.

We have to identify the fundamental barriers to alternate energy production that limit the performance or keep the cost too high or create environmental barriers. We must find ways to focus on those, rather than on demonstration projects. It's not the time to rush these technologies into commercial use too soon in an uneconomic way. It's time to step back and say let's parse this system. Let's look at what it would look like. Let's identify the barriers, and let's go to work and research in those areas that limit use. That's what I think, but it's not just the performance side—it's cost, environmental acceptability, and safety.

Tom Brownscombe, Shell Chemical: I want to make one comment about the synthetic fuels. We had a similar experience, except we have commercialized the synthetic fuels process and are building another synfuel plant. But I wanted to ask you a question—does global warming prove itself to be a serious issue in your view, and why should we take precautionary action?

Brian Flannery: I believe your remarks refer to Shell's gas-to-liquids technology, a 1990s development. I was referring to technology to make liquid fuels from oil shales, an effort in the 1970s and 1980s. When folks speak of precautionary actions to address climate change, such actions should include a wide range of steps, research being one of them. Looking at and affecting technologies, and discovering what their barriers are is real action. It requires real resources. It requires real prioritization and thought. It's not "no action."

I think all companies are taking action to become more efficient. For example, ExxonMobil has more co-generation capacity than any other oil company in the world. We produce over 2,000 megawatts. We didn't need credit for an early action to do that. It makes great economic sense, so we do it. As soon as the regulatory and enabling conditions are in place, we put it in place.

With strong management systems and disciplined investment efficiency, steps are easy to implement. We are also undertaking research with General Motors and with Toyota on advanced vehicles, including hybrid and fuel cell powered automobiles. However, fuel cell powered vehicles cannot be rushed into widespread commercial use. They are not economic today. But performing the research to create economic options is real and tangible action.

4

Opportunities for Carbon Control in the Electric Power Industry

John C. Stringer
Electric Power Research Institute

I begin by discussing the use of the new technique of roadmapping for the identification of longer-range technical challenges and illustrate some of the conclusions reached by the Electric Power Research Institute's (EPRI) *Electric Technology Roadmap: 1999 Summary and Synthesis* that are relevant to the topic we are considering. I want to look first at the global implications for carbon management and then consider some of the issues and the current options for the United States.

The key issue relates to energy and the role that it plays for human societies. At the most primitive level, the energy available to an individual was his (or her) own strength; with the development of family structures, this could be managed better, but the limits were much the same: the objective was basic survival. The first major change was the domestication of animals, such as the ox or the horse. The total energy used by individuals in the advanced societies is, by comparison, enormous, and the margin above the "survival minimum" can be used to achieve what we think of as "quality of life."

In the world as a whole, there are people presently living with energy availability across this complete range. In an early analysis, Chauncey Starr, EPRI's founder, distinguished four ranges: (1) survival, (2) basic quality of life (literacy, life expectancy, sanitation, infant mortality, physical security, social security); (3) amenities (education, recreation, the environment, intergenerational investment); and (4) international collaboration (global peace, global investment, global technologies, global R&D) Starr, C. 1997). Each of these related to a range of energy availability per capita and wealth production, as measured by gross domestic product (GDP) per capita. On that basis, the EPRI roadmap suggests that a global objective should be to ensure that the energy available to each individual is a minimum that corresponds to a level between the second and third of these classifications.

There is the issue of how this energy can be made available to individuals. As recently as 1950, electricity represented 15% of the world's energy usage; by 2000, this had risen to about 38%; and extrapolations suggest that by 2050, electricity will represent 70% of the energy use. The EPRI roadmap suggests that the target should be providing a minimum electricity supply of 1,000 kWh per person per year by 2050.

At the same time, there has been a progressive improvement in the efficiency of energy use. A common unit for energy use is "tonnes of oil equivalent" (toe), and the overall efficiency of use was determined by the GDP. In 1950, 0.35 toe was required for each $1,000 GDP (1990 U.S. dollars). By 2000 this had fallen to 0.31, and extrapolation suggests that by 2050 it could be in the range of 0.12 - 0.18. This quantity is called the "energy intensity," and for several years it has been decreasing at a rate of 1% per annum. EPRI's roadmap proposes a target of 2% per year.

The next issue is global population. This last year, the world's population exceeded 6 billion. By 2050, extrapolations suggest that this might rise as high as 10 billion, although earlier chapters have suggested that the most recent estimates may be somewhat less than this.

When these numbers are combined and the retirement of most of the world's current generating capacity by 2050 is considered, this goal is equivalent to adding 10,000 GW of generating capacity. This means building 200,000 MW of capacity per year, which at current costs represents investing something like $100 billion to $150 billion per year. While this is undoubtedly a large sum, it is less than 0.3% of the world GDP, and as EPRI's president, Kurt Yeager, says, "It is less than the world currently spends on cigarettes!"

The global efficiency of the production of electricity from the current fuel mix averages about 32%, which the EPRI roadmap proposes should be increased to 50% by 2050. Another important consideration is the "capacity factor" of a generating plant—that is, the fraction of the time that a given plant is in fact generating electricity. The overall global average is 50% for central station generation. The EPRI roadmap proposes that this be increased to 70% by 2050. However, further careful evaluation needs to be done to ensure that the manufacturing capabilities exist to meet these demands.

This gives an idea of the magnitude of the problem facing us over the next 50 years. In this chapter, I talk only about generation of electricity. I do not discuss the problem of delivery from the point of generation to the final user, although this too is a major issue.

Now, from the point of view of the workshop, the question is, How do we generate this electricity, and how does this contribute to the present and future production of anthropogenic greenhouse gases, specifically CO_2?

Let us review the situation in the United States. In the United States, the carbon emissions in 1995 were 524 MtC (million tonnes of carbon equivalent) for buildings (heating, lighting, and so forth), 630 MtC for industry, and 473 MtC for transportation. Essentially all of the transportation emissions came from petroleum, while 123 MtC of the buildings' emissions came from natural gas, 42 MtC from petroleum, and 355 MtC from electricity. For the industry total, 177 MtC came from electricity and the remainder from a variety of sources. In terms of primary fuels, the numbers were 628 MtC from petroleum, 319 MtC from natural gas, and 533 MtC from coal. As a first approximation, therefore, the three major categories made equal contributions to carbon emissions.

For transportation, the sources of CO_2 are many small, widely dispersed, and mobile entities. They need a storable, high-energy-density fuel. Petroleum-derived fuels fit these requirements very well. Removal of the CO_2 emissions from internal combustion engine exhausts will present a significant problem, and the costs are likely to be socially and economically unacceptable. In the longer range, hybrid automobiles, which are now being introduced, may help. Electric vehicles might have the effect of transferring CO_2 production from the vehicle to an electric utility generator. Fuel cells, particularly with hydrogen fuel, are the ultimate goal, but they are a few years away.

The most easily addressable source of CO_2 is from the generation of electric power, since there are a much smaller number of very large stationary sources. That is the primary topic of this chapter. As pointed out above, one factor of importance here is the increasing "electrification" of primary energy sources with time, and this pattern is reflected in the rest of the world. For example, in much of the world, mass transportation systems are increasingly powered by electricity, and (as indicated above) recent research has been addressing the electrification of personal transportation, although there are still significant barriers to achieving systems that are acceptable to the public.

The consequences of this scenario for meeting the carbon management goals also need careful study, and this has been done. A major problem is that much of the building of generating plant will take place in developing countries with large populations, notably China and India. The principal indigenous fuel available is coal, and it appears at the moment, at least, that there is relatively little natural gas available.

CARBON MANAGEMENT FOR THE U.S. ELECTRIC UTILITY INDUSTRY

As indicated above, electricity generation represents only about one-third of the anthropogenic CO_2 generation in the United States, and the complete removal of all this will be insufficient to achieve what is believed to be the necessary reduction. Nevertheless, for a number of reasons—most obviously the large stationary sources—this is likely to be the area in which a reduction is first demanded. As a consequence, both the U.S. Department of Energy (DOE) and EPRI have been giving the problem much thought. In general, the approaches may be categorized as follows:

1. Improvement of efficiency in electricity generation from fossil fuel-fired thermal systems;
2. Increase in the hydrogen-to-carbon ratio of the fuels used, with an end point of hydrogen as the fuel;
3. Increased use of so-called neutral fuels—those that depend on combustion in heat engines, but are regenerable: wood is an example, but various biomass fuels and municipal solid wastes are also examples;
4. Switching to non-heat engine methods of deriving energy from the oxidation of fuels to avoid the Carnot limit—for example, fuel cells;
5. Use of noncombustion heat sources for heat engines—nuclear fission, geothermal heat, solar heat;
6. Use of photoelectrically produced electricity—thermoelectricity is also a possibility of this type, given two adjacent locations of significantly different temperatures;
7. Increased use of hydroelectric generation, including low-head hydro;
8. Generation depending on the management of tidal variations;
9. Generation systems depending on the harnessing of ocean waves; and
10. Use of wind energy, for example, via wind turbines.

The last six of these are often referred to as "renewables," although this term is seldom if ever used for nuclear fission, while it almost always is for geothermal energy. It is common to treat biomass as renewable as well, although this could be argued. Hydroelectric generation, particularly from large reservoir systems is regarded as renewable in one sense, but environmentally damaging in another, and similar objections have been raised for large tidal schemes.

Improvement in the efficiency of end use of electricity clearly can also make very significant impacts on the emissions of greenhouse gases. However, as the recent rise in the sales of sport utility vehicles has shown, individuals are generally unwilling to sacrifice perceived personal benefits for something that may be for the greater good. This brings up an important point. In discussing the management of carbon, it is naïve to neglect wider issues that are more often thought of as sociological or political. For example, Congress has been unwilling to ratify the Kyoto Protocol because of concern about the impact on the U.S. economy and the negative effect it might have on U.S. industry's competitiveness in world markets. Significant impacts on the cost of electricity caused by carbon management might be unacceptable, although interestingly the public accepted significant increases in costs resulting from the introduction of SO_x and NO_x controls in the 1980s.

Over the last 100 years, the carbon intensity of world primary energy has been falling at an approximately linear rate, from 1 tonne of carbon per toe in 1900 to 0.7 tonne in 1990 (National Academy of Engineering, 1997). Extrapolation suggests a value of 0.55 by 2050. This rate of decarbonization is 0.3% per year, and EPRI has proposed a target of increasing this to 1.0% per year by 2030 and maintaining that rate thereafter.

This reduction was achieved largely as a result of switching to primary fuels with a higher hydrogen-to-carbon (H/C) ratio and of increases in hydroelectric generation in the early years and nuclear fission in the latter part of the last century. Both of these resources have essentially reached saturation in the United States.

In the United States for many years (certainly since 1920), coal has been the fuel that produces a little more than 50% of the electricity, and this has held steady as the generation capacity increased. In 1996, coal accounted for $1,797 \times 10^9$ kWh, or 52% of the total electricity generation, and this figure is predicted to increase to $2,304 \times 10^9$ kWh by 2020. In each of the last three years, the coal consumed by the electric utilities has been close to 900 million tonnes (for an example, see EIA November 2000).

For a time, oil was an important utility fuel, but at present, very little is used in this way. The use of natural gas has been increasing recently as new large, high-efficiency combustion turbines have become available, and this switch has been responsible for the continued increase in the overall H/C ratio.

It is worth reiterating that much of the expected increase in the demand for electric power over the next half-century will come from the developing world, and the greatest demand in the first few years will be from India and China. The largest internal fuel resource in both of these countries is coal.

The EPRI roadmap recommendations are aimed at improvements within the next 20 years, with some views extending out to 2050. DOE's Vision 21 scenarios are somewhat similar (DOE, 1999). It would be fair to say that within these time scales it is not believed that the so-called renewables will make much of an impact on U.S. electricity generation. Even the most optimistic predictions suggest less than 20% by 2050. The situation for nuclear power is much less clear. Recently, renewal licenses have been granted for a number of the older nuclear stations, extending their use for a further 20 years. If this continues, it seems probable that the nuclear contribution will be maintained for the next few years. It should be remembered that the proportion will fall as the total generation increases; presently it is about 20%. As yet, the climate of public opinion does not appear to have reached a point where new nuclear plants would be acceptable. However, it should be said that this would be the easiest way for the United States to reach compliance with Kyoto at minimum economic impact.

In leaving this aside, the issue becomes one of how to manage carbon emissions with a generation fleet that will consist of increasingly fossil fuel-fired thermal stations. By far the largest part of this in the past has come from large coal-fired Rankine-cycle plants, in which coal is pulverized and pneumatically injected into a large combustion chamber where it is burned. The burners are designed to produce a stable fireball in the center of the combustion chamber, whose walls are constructed of vertical tubes at the bottom of which preheated water is introduced. The water rises in the tubes, reaching boiling point at a point a little above the burner level; the heat is transmitted from the fireball by radiation. The cooling combustion gases then pass over a succession of further banks of tubes, superheating the steam. This is then expanded through a high-pressure steam turbine and returned to the boiler to be reheated. The reheated steam is then further expanded through subsequent stages of the steam turbine and finally condensed in a large condenser cooled by water from a large local source—the sea or a river, for example. As in any heat engine, the overall efficiency is determined by the difference between the maximum temperature and the minimum temperature of the working fluid. For the temperatures attained in a conventional large steam plant (538°C or so), the overall efficiency of a Rankine cycle is about 38% (coal pile to busbar). Efficiencies as high as 41% were achieved as long ago as the 1950s, but the maximum temperatures (650°C) were avoided because of materials problems. Research is currently in progress all over the world to attain higher-efficiency Rankine-cycle plants, but it seems unlikely that efficiencies much greater than 45% will be attainable in the immediate future.

Recently, essentially all of the generating plants that have been ordered have been advanced high-efficiency combustion turbines fired with natural gas. Many of these are "combined-cycle" machines, in which the hot exhaust from the combustion turbine (Brayton cycle) enters a heat recovery steam generator (HRSG); the steam from this is expanded through a steam turbine (Rankine cycle). Largely as a result of research funded by DOE's Advanced Turbine Systems (ATS) program, the newest generating systems will achieve an overall thermal efficiency of 60%. The largest of these systems generate approximately 400Mw(e) (megawatts of electrical power), which is the preferred size in the modern utility environment in the United States. In addition, they are relatively cheap to build, about half the cost of a fossil-fired steam generating plant of similar capacity, and the construction time is relatively short. However, they are natural gas fired; this fuel currently costs significantly more than coal on an energy basis, and there is some concern that a significant increase in demand may result in a further increase in price.

For many years, both DOE and EPRI have been conducting research into the advanced gasification of coal, and the use of the product in current-generation gas turbines has been demonstrated at relatively large scale. This technology is called integrated gasification combine cycle (IGCC) and is thus available as an option if either the price of natural gas rises too high or the availability is insufficient.

To give an idea of the magnitude of the problem, it is worth quoting an analysis done by EPRI a few years ago of a hypothetical Rankine-cycle generator located in Kenosha, Wisconsin, that was burning Appalachian bituminous coal (EPRI 1991). The coal contains 71.3% carbon, 6.0% moisture, 9.1% ash, 4.8% hydrogen, 4.8% oxygen, 2.6% sulfur, and 1.4% nitrogen (by weight ultimate analysis). The heating value is 13,100 British thermal units per pound (30,470 kJ/kg). The net efficiency of the unit (from the coal to the electricity delivered to the system busbar) is 35%. The corresponding coal burn rate is approximately 125 tonnes per hour. Typically, boilers operate with approximately 5% excess oxygen, and the flue gas characteristics are shown in Table 4.1.

From Table 4.1 it can be seen that the quantities of CO_2 are very large; if these were to be 100% sequestered as calcium carbonate ($CaCO_3$), this would represent 666 tonnes of product per hour, or close to 16,000 tonnes per day. The cation source would amount to the equivalent of 9,000 tonnes per day. This compares with a coal supply of approximately 3,000 tonnes per day and an ash production of 270

TABLE 4.1 Flue Gas Characteristics (principal components only)

Component	Tonnes per hour
O_2	93
CO_2	290
SO_2	6
H_2O	73
N_2/NO_x	1,134

tonnes per day. The masses involved more than 50 times greater than those involved in flue gas desulfurization, which is a practice used in essentially all U.S. boilers firing high-sulfur coal. It should also be noted that the CO_2 represents about 18% by weight of the exhaust gas and 15% by volume, which emphasizes its relatively dilute character. Since this "typical" plant produces 2.32 tonnes of CO_2 per tonne of coal, the utility industry is currently producing 2.1×10^9 tonnes of CO_2 per year from burning coal. This does not take account of the CO_2 produced from units burning natural gas.

U.S. coals have sulfur contents that range from as low as 1% to more than 4.5%, and environmental regulations require that this be removed from the combustion gases before they are released from the stack into the atmosphere. This is done by fuel gas desulfurization (FGD) systems, which are located between the boiler exhaust and the stack. Since the regulating legislation was passed in the 1970s, there has been considerable research and development into FGD systems, and more than 30 are now commercially available. The majority are "wet" systems and depend on contacting the exhaust gas with an aqueous slurry of calcium oxide, for example the product is calcium sulfate, which is environmentally benign and (as gypsum) has some commercial value. The product slurry typically goes to settling ponds and is then trucked away. Some dry systems are available for plants in regions that are relatively arid, but the basic chemistry is the same.

The utility industry is also familiar with satisfying environmental legislation that limits the emission of oxides of nitrogen (NO_x) from the plant. This can be done either by modifying the combustion process (low-NO_x burners) or by postcombustion techniques, such as selective catalytic reduction.

Particulates are also removed from the combustion gas, by techniques such as electrostatic precipitation (ESP) or the use of baghouse filters.

The point here is to indicate that physical and chemical methods of removing contaminants of various kinds are well known in the industry, and this involves the treatment of the entire exhaust gas stream.

CARBON DIOXIDE CAPTURE AND SEQUESTRATION

This leads to the other major option in carbon management. As indicated above, in the United States the utility industry is the prime target for carbon control. Our first priority is to address the control of emissions from these units because we rely on coal-fired Rankine steam plants for more than half of our electricity, and because there seems little prospect of materially reducing this for some years. The coal-fired, relatively low-thermal-efficiency units are the most significant CO_2 emitters in the utility system. Control of emissions from these units involves the separation of CO_2 from what is a relatively dilute exhaust gas, its *capture* in some way, and finally its disposal in some environmentally acceptable and long-lived manner. This last step is called sequestration.

There have been several meetings over the last few years addressing this approach. In particular, the U.S. Department of Energy recently summarized the issues in *Carbon Sequestration: Research and Development* (DOE, 1999).

The separation of CO_2 from gas streams is reasonably well known. The most obvious example is its removal from natural gas. In some cases, the amounts may be quite large: some economically recoverable natural gas reservoirs contain significant amounts of carbon dioxide. The Sleipner West field in the North Sea (for example) contains 10% by volume CO_2; the sales specification is not more than 2.5%. Statoil (the Norwegian state oil company), which operates the field, uses an amine solvent technique to separate the excess, which is then pumped into a reservoir 1 km below the seabed. Approximately 1 million tonnes of CO_2 is separated annually, which is about 40% of the model Kenosha plant described above.

Sleipner West illustrates one aspect of the current approaches to managing CO_2 in coal-fired fossil systems. The favored approach is to somehow generate a CO_2 stream that is highly concentrated. The DOE report mentioned above lists the following methods:

- Chemical and physical absorption,
- Physical and chemical adsorption,
- Low-temperature distillation, and
- Gas separation membranes.

Chemical absorption is preferred to physical absorption for low to moderate CO_2 partial pressures, which is the case for fossil power plant exhausts; typical reagents are alkanolamines such as monoethanolamine (MEA). However, these have to be regenerated using hot steam stripping to produce the high-concentration CO_2 stream, and various studies have shown that this regeneration imposes a significant economic penalty.

Adsorption processes depend on materials having very high specific surface areas and a high selectivity for the target gas. Zeolites are naturally occurring examples of such materials. An International Energy Agency (IEA, 1998) study concluded that this approach was unlikely to be economically viable for power plants, but the DOE report notes that adsorption techniques have been used in some large commercial CO_2 point sources in hydrogen production and natural gas cleanup systems.

Low-temperature distillation is widely used commercially to produce high-purity CO_2 from high-concentration sources. It does not appear to be appropriate for power plant exhaust gas treatment.

Gas separation membranes have not been used to any great extent thus far for CO_2 separation, but there has been a great deal of interest in the development of gas separation membranes in recent years, and this may well be an area meriting further research.

There are some interesting variations on addressing this overall problem that could change some of these conclusions. For example, the use of an oxygen-blown IGCC approach will produce a high-concentration CO_2 gas stream since there is no nitrogen diluent. In addition, the sulfur and fuel-bound nitrogen species can be removed in the gasification process. Adding a shift reaction further increases the CO_2 generated within the fuel preparation reactions preceding combustion. Furthermore, the combined-cycle aspect of the process can lead to an improvement in overall cycle efficiency.

Having produced, by whatever means, a high-concentration CO_2 stream from the fossil fuel-fired plant, the issue moves to sequestration. Again, the most common approach at the moment is to compress the gas to form a liquid, which can then be pumped through pipelines to a sequestration site. Such sites include the following:

1. Geological sites
   - Deep porous strata
   - Deep saline aquifers
   - Freshwater aquifers unconnected to potential drinking sources
   - Spent oil wells
   - Depleted natural gas wells
   - Deep unminable coal seams
2. Marine sites
   - Very deep regions
   - Shallower regions that favor carbon dioxide hydrate formation
   - Near surface regions that allow biological capture processes

All of these sites have been discussed in considerable detail over the last three to five years in a number of symposia. The DOE report referred to above summarizes a number of them. The major issues associated with this concentration-pumping transfer-sequestration scenario are the following:

- The economics of the concentration step from the very large volume of relatively dilute exhaust gas from a utility boiler,
- The problems associated with the distance over which CO_2 will have to be pumped between the generation site (the power plant) and the sequestration site,
- The risks associated with leaks in the transport lines (since, unlike natural gas, released CO_2 will concentrate in low-lying regions adjacent to the leak and is not easy to detect),
- Leaks from the geological repositories,
- The ultimate capacity of geological repositories, and
- Local environmental effects on marine repositories.

Many authorities believe that these issues are not insurmountable from a technical point of view, but most also agree that licensing and insurance issues in the near term may present problems.

There is a clear preference for very long-lived or even permanent sequestration. Such permanent sequestration is offered by an examination of geological processes, since it is clear that over archaeological time scales, a considerable amount of CO_2 has been sequestered—for example, as oolitic limestone deposits and dolomite deposits. Nature has sequestered carbon in two ways:

1. As calcium carbonate generated by marine animals of various kinds, principally as shells or exoskeletons that are deposited on seabeds—this appears to be the method of formation of the very extensive oolitic limestone beds; and
2. As carbonates generated by silicate → carbonate exchange, in which the by-product is silica (SiO_2)—this is often referred to as the weathering process.

It has been suggested that some of the geological sequestration routes described above will eventually lead to permanent sequestration through the second of these paths. The rate of the exchange reaction is currently unknown, and the possibility of accelerating it, either by catalysis or by the use of high-surface-area silicate materials, has been discussed but not studied. This would seem to be a fruitful area for further research. There has, of course, been a significant discussion of the first option, in the biomimetic approach proposed by Bond and coworkers (New Mexico Institute of Technology, Personal Communication). Another approach that has been proposed is to catalyze the ocean processes responsible for the near-surface capture of CO_2, for example, by using iron salts spread on the surface.

While the benign and more-or-less permanent sequestration offered by these techniques is attractive, as is the potential of being able to eliminate the concentration and pumping steps, the mass flows are still daunting. For example, if CO_2 were to be sequestered by the pumping of seawater, with a calcium ion concentration in the seawater of 400 g/tonne, through a separation vessel at the utility site, 100% removal from a unit of the size of the model Kenosha plant would require a flow of 18 million tonnes of seawater per day. This seems like a very large number, and indeed it is, but the cooling water flow through such a unit would be of the order of 2.5 million tonnes per day. (Note that other methods for capturing the CO_2 as carbonates are also being considered, and it is by no means clear that such a large flow through the plant is unavoidable!)

CONCLUSIONS

The important point in all this is that the problem is very large, whatever method is chosen to achieve a solution. There is no clearly superior method available at this time, and careful and thorough research is necessary for all the candidates that have so far been proposed. The hunt for as yet undiscovered approaches is an important—perhaps critical—part of the necessary research.

It is quite important not to commit too early to a process that, after some experience, turns out to be unsuitable or to write legislative goals that cannot be attained at a reasonable cost.

This is obviously an argument for more research, but it is also important to understand that in all aspects of this complex issue, the clock is running. Planning the research that is needed, specifying the goals that must be achieved, and deciding the times by which answers must be developed are essential. There is a large stakeholder group that includes in one sense everybody! Without good understanding by the stakeholders, acceptance of the limitations and the consequences will not be possible.

Roadmapping is, in my view, the only planning technique available to us that can develop a research approach appropriate to the problem.

REFERENCES

Bond, G.M. New Mexico Institute of Technology. Personal Communication.
Department of Energy. 1999. Carbon Sequestration: Research and Development. U.S. Department of Energy Report DOE/SC/FE-1. Washington, D.C.: U.S. Government Printing Office.
Electric Power Research Institute. 1999. *Electric Technology Roadings*:C1-112677-V1
Electric Power Research Institute. 1991. *Economic Evaluation of Flue Gas Desulfurization Systmes, Volume 1.* EPRI Final Report on Research Project 1610-6, Report No. GS-7193. Palo Alto, Calif.
International Energy Agency (IEA). 1998. *Carbon dioxide capture from power stations.* Available at www.ieagreen.org.uk/sr2p.htm.
National Academy of Engineering (NAE). 1997. *Technological Trajectories and the Human Environment.* Washington, D.C.: National Academy Press.
Starr, C. 1997. Sustaining the human environment: the next two hundred years. Pp. in Technological Trajectories and the Human Environment. Washington, D.C.: National Academy Press.

DISCUSSION

Jim Spearot, General Motors: Given your concerns regarding the licensing of large, centralized power generating stations and plants, what are your thoughts on a distributed electrical production system using fuel cells, and perhaps natural gas, as a way to get around the additional needs for electricity.

John Stringer: Thank you. I have a slide on this that I didn't put in. The issue of distributed power is one that has been exercising us for quite some time. We feel that distributed power does have a place. Exactly what that place is, is quite difficult to calculate. In very simple terms, it depends on whether in our jargon, you can "snip the wires."

By this I mean that if somebody has a small power plant, a distributed power plant for say a mall, there are two possibilities. In the first, they rely entirely on their own power plant for their needs and do not require a utility to provide a backup in the event that their own unit goes down. In the second, which is much more common, they require the utility to put a line in and be available to them as a backup. The overall costs of the two are very different. The installation of the line is a cost item, of course, but the most significant item is the reserves that the utility has to maintain—unused mostly—to be able to supply them in the event they need it. This represents a capital cost to the utility that generally is not earning any money and for a large entity may be substantial. If the utility requires them to pay for what amounts to a rental for this backup, then the economic advantages of the distributed system may well disappear. That's one issue, and there are some other issues connected with distributed generation, but nevertheless, it's an excellent question, and we are looking at it very carefully.

Alex Bell, University of California at Berkeley: In separate parts of your talk, you discussed the desirability of burning a fuel with pure oxygen and then scrubbing out the CO_2, as opposed to burning the same fuel in air prior to carrying out the removal of CO_2. Can you talk about the trade-off in energy? Is there any net energy efficiency in large-scale air separation, where you use oxygen to burn your fuel and then remove CO_2 from a nitrogen-free exhaust gas?

John Stringer: Again, it's a good question. For the calculations that we have done so far, with the separation techniques that we have available to us at the moment, the answer is yes. This calculation is quite difficult. For a coal-burning Rankine plant, the figures I gave show that the actual exhaust is quite dilute in CO_2—around 18%—and consequently, a concentration process involves moving a considerable volume of gas. We and others have done detailed calculations for the widely used amine solvent technique for removing CO_2 from a gas stream that was used at Sleipner West, for example, and it doesn't make economic sense at the moment. The costs aren't wholly unreasonable, but it isn't all that obvious that they can be reduced sufficiently to make the process economic for treating exhaust gas.

Alex Bell: Can you quote a figure in terms of the percentage of heating value of your fuel that goes into CO_2 separation by one strategy versus another, because that would put everything on a common footing?

John Stringer: For some of these things we have the numbers. The calculations for the amine stripping were done by the International Energy Agency a couple of years ago. I don't have the numbers at my fingertips, but we have redone the numbers and we come up with something that is about the same. However, other people also have done the calculations, and these are quoted in the DOE report that I mentioned. As I recall, their numbers are slightly more optimistic than the IEA results, but I don't know what differences led to that. There are numbers available, and there are a number of studies, largely funded by DOE, that are currently examining the economic issues. EPRI has a number of industry-accepted methods for doing economic analyses of this kind that have been applied to, for example, fuel gas desulfurization techniques.

George Helz, University of Maryland: You mentioned a figure that I have often seen. It's something like 38% efficiency in a typical plant—that is, within the plant—from the coal entering the firebox to the wire carrying electricity away from the plant. I have often wondered what the total efficiency is from the coal deposit at the mine to the heat in the consumer's toaster. Do you have such figures? Are they largely different from the 38%? Is there a lot of additional energy cost in the total stream?

John Stringer: The coal prices in the United States are fairly modest; consequently, the costs incurred in the transfer from coal mine to the plant are not really very large. In terms of efficiency, there isn't any significant penalty at that stage—the total energy loss in the transportation is really quite small. I haven't seen a calculation, but it would be very easy to do since transportation within the United States is by train or barge. Also, on the transmission losses, in the United States. Once again, transmission losses are, at the moment at least, fairly small. I think the question is interesting, and I should be able to express efficiencies in those terms. When I get back to EPRI, I will try to put something together.

Jack Solomon, Praxair: This is a partial answer to Alex's question. For the amine CO_2 separation systems, the extra CO_2 required is 30-50%. The penalty for an oxygen-burning plant, just to look at that, is 15-20%. Thus, the oxygen-burning plant is a little better. It depends a lot on the details.

John Stringer: It depends, for example, on whether you've got a market for the nitrogen.

Jack Solomon: Also, you have to do something with the CO_2. You mentioned integrated gasification combine cycle, and you mentioned penalties on the reliability. Do you have any comments on what those penalties are?

John Stringer: The experience base is, at this point, fairly small. I'm thinking mainly of the plant in Holland, where they sought a high degree of integration. The level of integration that has been required for this plant, coupled with the problems with the combustion turbine, caused it to be listed in the press as one of the major economic disasters in the Netherlands over the last 5-10 years. So the consequences were enormous there.

Now the fact is that you don't need that level of integration. The marginal benefits you get from full integration are relatively small. If you don't integrate, then the reliabilities are determined by the least reliable components of the plant.

Brian Flannery, Exxon Mobil: I understand that Finland is seeking a license to build a new nuclear power plant, partly in the context of what Finland has to do in terms of climate change. In the United States, we are going through relicensing. What are the prospects for nuclear energy in the United States? That is, what will the scale of nuclear be in the U.S. utility mix say in 10 years, and what are the prospects for having more nuclear?

John Stringer: We have slightly more than 100 nuclear plants at the moment, which are producing approximately 15-20% of the electricity in the United States. The first three relicensing cases have all gone through. We have another two or three in the works at the moment, and we have every reason to suppose that they will go through as well. So consequently, I don't expect that we will lose many nuclear plants in 10 year's time. We might lose one or two for basic reasons like the New England one, but I expect it to be roughly the same. The reason there is a falling percentage of total generation is because of the growth in generation we expect over that time.

Will we build a new nuclear plant in the next 10 years? There's not a chance in the United States, I believe. In the next 20 years, maybe. It depends a lot on what is going to happen with global warming, I think, because once it gets to that point, I think people will be looking at all the possibilities, and one possibility is nuclear.

What will actually happen depends on the outcomes of cases currently going through the courts in which two or three big utilities are suing the federal government for not actually completing its part of the agreement to store the waste fuel. This issue won't be resolved in the near future, and of course it depends very much on not having another nuclear disaster.

Fred Fendt, Rohm and Haas: It has been maybe a quarter of a century since we've built a new nuclear plant in the United States. Has research continued in nuclear power plants in that time? Would a new nuclear plant be markedly different from any of the existing ones?

John Stringer: We had a lot of research done on conceptual systems—when I say research done, I mean up to the stage of sort of fairly detailed planning studies. This has followed two major directions. One of them is the use of what's called a "pebble bed" reactor, and that's a higher efficiency type of reactor. The Germans played a large part in that research, and of course we had a program going in the United States. The South Africans, I believe are looking at the possibility of a pebble bed reactor at the moment as well.

The other direction is building what we call an "advanced light-water reactor," which is designed to be able to achieve a sensible and controlled shutdown in the event of a component failure, such as a circulation pump failure. We have all of the designs for this. It was funded by EPRI and by the Department of Energy. So I think the next reactor we build, whenever it is, is not going to look the same as the reactors we have at the moment.

Geraldine Cox, EUROTECH: When we're in a situation like this, we often have to examine the original paradigm under which we operate. In this case, it is the supply-demand of energy. With coal plants, I recognize the size of the resource and the need to continue its exploitation. Yet coal is very inefficient in the sense that it is not able to supply a just-in-time response to energy demand, where gas turbines and others can turn on and off much more efficiently in response to peak demand. A cold start in a coal plant to full operation takes several days. Is industry studying approaches to minimizing off-peak power production to store the energy in some way that it can be used more efficiently during peak periods?

John Stringer: Yes, it is, but I have got to tell you that the scenarios do not make an enormous difference. It's at the margin.

First of all, the overall generation pattern required by any utility consists of essentially a base load, together with a number of fluctuating demands with different time scales. There are some variations on a weekly time schedule. There are ones with six-month variations, there are daily fluctuations, and there are variations that are of much shorter times than that.

A popular television program comes on at let's say 7:30 p.m., and the demand just rockets up. So what we do is we always have a generation mix, because if you have something that has been designed to satisfy the base load, it can operate best at the design point. Then there are some that are peakers—for example, a simple combustion turbine, because it can turn on very fast.

However, if we go to the high-efficiency combustion turbine systems that I was telling you about—the combined-cycle ones—their start-up time is not fast. We can perhaps start them up in simple cycles quickly and then bring up the steam generator, the Rankine part of the cycle, over a longer period. However, they will be operating only at around 35% efficiency or so when they are operating in simple cycle mode. So, yes, we do have a mix, and there is a quite complicated dance that goes on about what you dispatch at any particular time.

The other question you had related to storage of energy not used. The best way is pump storage. What you need is two lakes that are around about 1,000 feet apart. There aren't too many of those. Where you have them, they present the best opportunity. The pumps operate wonderfully. The hydro-generator works great, and it's very high efficiency. There is a plant in the United Kingdom in a place called Dinorwick in Wales that has been operating like that for many years.

We have put a compressed air storage plant in Alabama, not far from Montgomery, that uses a salt cavern. Essentially, there are big salt masses down there, and you can open up a big cavity by dissolving the salt by injecting water. At the surface, there is a gas turbine-driven generator, which also drives an air compressor, and you store the compressed air in the cavity. When the demand for electric power increases, the stored compressed air is expanded through the compressor, which acts as a supplementary power generation turbine. Actually, the operation is a little bit more complicated than that, but at least this shows the storage principle.

Heinz Heinemann, Lawrence Livermore National Laboratory: You extrapolated from the year 2000 through the year 2050 and showed an increase in excess energy demand from 38 to 70%. Does this take into consideration the limitations in transmission of energy from production to consumers?

John Stringer: That's an excellent point. The most sensitive part of the overall electricity production and supply system in the United States, at the moment anyway, is the transmission system and, to some extent, also the distribution system. These are jargon words. Transmission means the long-range transfer of electricity from the generator to the main transformer park. Then distribution is from these transformers to the final user.

Some components of the distribution system have been in place for a long time. In New York, for example, I am told that we are still using part of the distribution system that was put in by Edison. So demand to ensure high reliability in the power delivery infrastructure is the number one priority in the industry in the United States at the moment. The other reason I didn't have it on my slides is because it's not really a carbon management thing, but in terms of the electricity supply industry, it's very important.

Klaus Lackner, Los Alamos National Laboratory: I have two comments. One is I heard a large number of discussions on the question of increased efficiency, but I really would like to come back to Jae Edmonds' comment that ultimately we have to go to zero emissions. If you go there, I don't think you have a choice but to deal with sequestration.

The other point deals with scrubbers. I agree with your number. I would even say that calcium oxide is not an adequate scrubber for the simple reason that calcium oxide started life as calcium carbonate. You could use magnesium silicate instead. Again, the transportation issue would indeed be overwhelming.

So the conclusion from this would be that you have to put the processing plant at the site where your scrubbing materials come from, because that is the largest mass flow in the system. The nice thing is that mineral carbonation is actually an exothermic reaction, so it works without requiring energy. This would allow for a different location for the power plant.

John Stringer: Yes, thank you. I had a slide, which I didn't show, that was going to comment on your work, because I think it's extremely good. We are reviewing these alkaline earth silicate deposits and going through the reactions that you have described, the classic weathering reaction I think, as part of our continuing attempts to imitate the ways in which nature has achieved the very long term sequestration of CO_2. This is an extremely attractive thing to do, and I think we need to do much more research in that particular area—for example, by studying the kinetics and looking for ways to accelerate the relevant reactions, but I think that's nice work of yours. I like it very much.

Glenn Crosby, Washington State University: I had a question for our last speaker regarding the research in new reactor types. Could you inform the audience concerning the advances that have been made, say, by the French, who are using nuclear generation of electricity much more predominantly than we are, into their reactor design and where they are in their reactor performance?

John Stringer: Their reactor designs are really not all that dissimilar to ours. What is different is in their overall plant: first of all, they have a much higher proportion of their electricity generated with nuclear plants, as you know. Again, most of their nuclear plants were built some time ago. There were a few differences, but there isn't an intrinsic difference.

It isn't like a Canada Deuterium Uranium (CANDU) plant for example, which is quite different, or the gas-cooled reactors that the British developed. The French use a light-water reactor. However, their overall plant philosophy was ultimately based on getting the breeders going to close the fuel cycle. As you may recall, they had the demo Phoenix plant, and they were planning to build the one that was going to be the first of the actual units to do this, sort of make the whole thing work together.

It was called the Super Phoenix. They had a major problem with that because it used a liquid sodium potassium coolant, and they had a coolant leak into the water. This caused a very significant problem.

I don't think the problem was insoluble, but it was sufficient at that time to cause serious political and social questions inside France about whether France wanted to go to the route of the breeder reaction. At this point, the whole thing is on hold. At this time, EDF (Electricité de France) is looking at coal power plants. It has a big circulating fluidized bed reactor built in France, and it is focusing on international marketing, so EDF is marketing as many or more fossil plants as nuclear power plants internationally. Nevertheless, the intrinsic nuclear experiment has been very good. EDF sells cheap power to everybody else around. So it's been successful.

Panel Discussion

Richard Foust, Northern Arizona University: One thing I haven't seen addressed in the presentations this morning has to do with the change in life-style. For example, in China they are moving from bicycles to scooters to automobiles. In projecting 50 years into the future, this may have a more significant impact on carbon emissions than population growth. Has this issue been dealt with in the models, and how do you do something like that?

James Edmonds, Pacific Northwest National Laboratory: The simple answer is yes, this is probably the biggest driver, as you rightly suggest. It's the largest force moving emissions upward. It is much more important than the simple numbers of people. It is the big-ticket item.

John Stringer, Electric Power Research Institute: I agree with that, and I might add one thing to it. We, and I think others, have looked at the growth of megacities—those with populations in excess of 50 million. Megacity growth leads to a possibility that carbon emissions may go back a little bit if the growth also incorporates a public transportation system.

Chandrakant Panchal, Argonne National Laboratory: This question concerns the cost and technical issues of CO_2 transportation. You have to properly link the coal and fuel, sources of CO_2, its production and distribution, and its users. CO_2 must be separated from diverse sources—utilities, refineries, and so forth. Most sources are in towns and cities. Many different cost estimates have been made. Does anyone want to comment on the reality of these, or are we embarking on something that is going to be very difficult to do?

David Thomas, BP-Amoco: Concerning the cost of CO_2 transportation, I have talked to people who build CO_2 pipelines, and they use some rather interesting units. They say the cost for a pipeline for high-pressure CO_2 (2,200 pounds per square inch or 150 atmospheres) is on the order of $15,000 to $40,000 per inch-mile of capacity. This means that a 10-inch pipeline is $150,000 to $400,000 per mile for construction costs.

In terms of carrying capacity, my recollection is that a 14-inch-capacity pipeline can carry on the order of 5 million tonnes of CO_2 per year. I may be low there. I'm using as an operational number for transportation for a couple hundred miles (300-400 km) an order of about $1 for a ton of CO_2. This is based on experiences in the Southwest, where they are moving CO_2 down pipelines for enhanced oil recovery. It's on that order, but the exact number depends on how far you transport it and under what kind of pipeline conditions. There are people who say that if you want to do a complete sequestration job using centralized storage in reservoirs of various kinds, you would effectively need to double the natural gas infrastructure. I think that's probably the high end. The low end is probably 10% of that. So somewhere in the middle lies reality.

Alan Wolsky, Argonne National Laboratory: I wonder if anybody has thought about how much we should spend on this? I don't mean how much does any single option cost. I mean, Is there some moral obligation to future generations to now raise the price of carbon by 10%? If we did this, or other things, would we be doing our duty to the future? At some point, I think one has to ask this to get a grip on the problem and its feasible solutions. I wonder if the panel would share its thoughts on this.

Brian Flannery, Exxon Mobil: I don't think people have asked the question quite the way you posed it, but the challenge is that no one knows what the cost of controlling carbon in the economy is. Kyoto was an attempt to control the amount of emissions at unknown cost. Others have suggested options such as putting a cap on the cost and requiring a permit to use carbon. Now, however, you are up against exactly the question you asked, What is the right level? How do you do it? Why should you do it? You are also up against the question, What will it achieve? Bill Nordhaus is an economist who has tried to show ways that you might impose a cost now in the hope of achieving something over a long period of time. Jae will be much more familiar with his numbers than I am, but the type of cost conclusion he came to was a rather low cost, compared with the cost of emissions-based outcomes that people have looked at so far. The answer to what is delivered depends on the technologies that are produced over the next 20 or 30 years and the climate signal that this level of cost might generate. Frankly, I think here, we're in assumption space. We don't really know the answers.

James Edmonds: I would like to add one more point. Given the stock nature of the carbon problem, economics is fairly familiar with this cost exercise, and the problem was solved about 70 years ago. The answer is that the cost of controlling carbon is not a fixed price. You start off with an initial value, and it should rise roughly at the rate of interest of the economy. If the rate of interest is roughly about 5%, controlling carbon costs should rise at about 5% per year. So if you start off at $10 a ton in the first year, it should be $10.50 in the second year, and so forth.

Alan Wolsky: Perhaps, the work of Harold Hotelling is in the minds of those trained in economics. His work concerned the desired rate of capital accumulation or nonrenewable resource depreciation. As I recall, his answer was the market interest rate. Consistent with his line of thought, it seems to me the comparison is between the wealth that you would otherwise pass onto future generations and the amount of money you spend now on passing on the same atmospheric conditions, and the interest rate has little do with how much we spend now. The interest rate has something to do with how much Hotelling's work suggests we should increase that spending over time. Perhaps our economic colleagues would share their views.

James Edmonds: Well, actually it has a lot to do with it. It's the rate at which you can transform wealth from one period to the next, so it's a very important determinant of any efficient solution to the problem. There are an infinite number of inefficient solutions, which impose either extra costs or additional costs borne by somebody else. It is an important element.

Alan Wolsky: So, should we spend $10 per ton or $50 per ton this year?

Richard Alkire: Maybe we should go to the next question. I should point out that we discussed how to shape the content of this Chemical Sciences Roundtable workshop and came to the conclusion that the one question we would use to try to do that is, If there is a carbon tax, what research in chemical science and engineering would be needed in order to understand how to proceed? So while there are many issues to the topic beyond the chemical sciences and research, the focus is what could be done by the chemical sciences community operating in this climate that would be helpful. Next question.

Dennis Lichtenberger, University of Arizona: We have heard some different comments on synthetic fuels and synthetic chemistry in general. A couple of people commented that they had negative experiences with the synthetic fuels issue. In contrast, another person mentioned that his company is now building a synthetic fuel plant and that synthetic gas chemistry is very important in many of its products. What is the feeling for the future role of the chemistry in this area?

Brian Flannery: The synthetic fuels program I was referring to was liquid fuels from coal and oil shales. Synthetic gas fuel is gas to liquid, which is a totally different technology. One of the reasons for that technology is to bring to market gas that otherwise doesn't have a market. The market that it serves is a totally different market than you would usually have for gas, because this process produces high-quality products that can be transformed into a premium liquid fuel, a diesel fuel, or a very interesting feedstock for a number of chemical processes, but that's a different market and regime from the efforts to provide massive amounts of liquid fuels from coal or shale. You are also consuming a lot of the energy when you transform gas to liquid. With coal you are working with a vast resource and converting it to a liquid, of which there was a smaller resource and problems with security.

So they are totally different technologies. The gas-to-liquid technology is an extremely attractive technology in certain options. The question is how to make it commercial. It does have interesting issues for climate change, because the emissions occur at the site of manufacture in a gas-to-liquid plant. It raises interesting questions about Annex 1 versus non-Annex 1 countries and where you would even site such a plant. The market may also raise interesting questions for corporations that have taken on internal carbon reduction targets, because emissions from these plants are operational emissions, not end-use emissions by the consumer.

Rosemarie Szostak, Department of the Army, Army Environmental Policy Institute: This question is for Brian, who had talked about government mandates as opposed to proactive behavior. We work under government mandate, not a proactive policy, although proactive would be nice. How would you envision implementing a proactive policy for carbon management that would be amenable to both the government and the general public?

John Stringer: It's an interesting question. Carbon mitigation will be similar to the experience we had with sulfur removal 20 years ago. At that time, the public was very keen to have sulfur removed. We went through the calculations and indicated that the sulfur removal to the levels that the government wanted to produce would increase the cost of electricity by about 35%. It was felt that at this level, the public desire to have sulfur removed was sufficiently high that it would in fact accept that. So legislation went through on that basis.

In fact, because of developments in technology, the actual cost impact now is quite a lot less than the 35% we had originally calculated. It's now probably down to around 15%. That is, the cost of electricity from coal is 15% higher than it would be if we didn't have to scrub the sulfur out.

Now I think the question with CO_2 mitigation is going to come down to a similar question. How much is the public concerned about CO_2 and global warming, and how much of an increase in the price of electricity is it prepared to accept?

Rosemarie Szostak: Turn that around. I'm asking, short of a carbon tax or mandating the control of carbon emissions, is there a way of being proactive that would be reasonable from a policy perspective? We certainly know from industry, and the BP representative who pointed out in his talk that you can decrease your emissions 10%, that it's good business. You save money. How do you come up with a policy that is not a government mandate?

John Stringer: We can't see a positive revenue side to removing CO_2. This was discussed quite a lot in the Department of Energy panel that I was on. Once again, nothing came through that looked like a reasonable market. We can possibly talk about this in more detail, about specific cases.

David Thomas: Let me add a little bit from our point of view. At the top of my list of targets was enhanced oil recovery, because that is the only use for which people are willing to buy CO_2 and pay us real money for it, so we can generate a revenue stream that offsets the cost of capture. With some modest changes to the enhanced oil recovery processes, one can turn them simultaneously into sequestration opportunities, which is the reason they are very attractive.

I think I agree very much with Brian that there needs to be an economic drive for most of these sequestration opportunities and emissions reductions to occur. We are of course pushing the cost reduction and savings through energy management. At the first stage, we are targeting those opportunities that will generate a revenue stream.

I don't like the idea of adding taxation. If anything, I would prefer a tax incentive. In other words, if I reduce my emissions, I would be incentivized. The guy who didn't do it would be taxed by paying a higher proportion of the income tax, rather than a punitive tax.

Brian Flannery: I think that in the first place, there are always some opportunities to reduce emissions through energy efficiency, fuel switching, things like that. The problem is that once you have installed the new advanced turbine this year, you are not going to do it again next year. It's a question of the rate at which you can afford to deploy these new technologies and the total reduction they deliver.

I think industry is reasonably good—or at least on the scale of multinational corporations—at having management systems in place that identify, plan, invest, and operate to improve efficiency. I know we have an energy management system that has been part of our system since the mid-1970s. It's been refurbished from time to time. I must say we find it perfectly capable of identifying attractive economic opportunities to reduce emissions. We don't find emissions trading adds any bit to our capability to make these types of decisions or investments.

There are perhaps ways to raise awareness about opportunities to reduce emissions that are proactive. Renewables are also potentially an opportunity. It's not one our corporation sees as attractive to shareholders. We've been there, done that, and lost a lot of money frankly, but there may be small opportunities for renewables in niche markets.

I think that when you are talking about sequestration and hydrogen, you are not talking about economic ways to move forward at this point. You are talking about demonstration projects, or research and development to make them less expensive, but today I think that's as far as you can go. They will be expensive, especially at this point. There is no way around this. It is a question of whether research and development can perhaps make them less expensive or options you want to deploy 30 years from now. However, I'll repeat, research is doing something. Doing it well and thinking about the systems and the infrastructure implications represents a major planning exercise.

Alex Bell, University of California at Berkeley: I want to pose a question for the general issue of setting the research agenda, and let me create a context for this question. Let's look at electricity generation, where it has been identified that you want to do air separation to produce oxygen. You want to have efficient burners for the efficient conversion of fuel to energy, efficient means of electricity generation, efficient scrubbing or removal of CO_2, and transportation of that CO_2 to the burial site, and then efficient burial and disposal of the CO_2.

So the questions are, Where are the long-term opportunities for research to be carried out by academics and researchers in the national laboratories? How do we use federal tax dollars efficiently for this purpose?

John Stringer: It's connected with the previous discussion on the last point. That is to say, What is the value of the eventual product? You can only spend money that in some way relates to the cost savings you expect. For sequestration, we really don't know what we are doing.

The things that you hear about—the Sleipner West project and the CO_2 reinjection into wells and so forth—are really very small things that don't in themselves permit scaling up and addressing the major problems. Now the major problem is permanent sequestration. There was a brief mention of permanent sequestration in the form of going through the silicate–to-carbonate reaction. You don't have to bother about anything, it is there forever. On the other hand, putting CO_2 into a hole in the ground is always a little bit disturbing, because what went down can come up. The experience with the injection of CO_2 into oil wells is just like that: 30% of it comes straight back.

So there is a lot of research to be done on let's say the fundamental aspects of the full train in the separation and sequestration. The things that we have at the moment are not good. In my view, they will not translate into practice.

James Edmonds: I might just add a couple of things to that. One is you want to have a fairly broad portfolio, and it needs to go all the way back to some pretty basic science, because if you are not laying down the basic science foundations for the next generation of technology, you are probably not going to get what you are looking for. That includes materials sciences and biological sciences. Then, as you move up toward the kind of things that the private sector tends to do best and you look at the technologies that are promising, which in many of the scenarios the concentrations are limited, you have a major role for technologies. Things such as the commercial biomass look to be areas where funding is at present pretty minimal and the marginal value of increasing support is fairly high. Similarly, fuel cells would fall into this category.

I think one of the big things that is going to have to happen if capture and sequestration is ever going to come to pass is that we are going to have to be able to guarantee that we know where the carbon is. If you think about it, any of the scenarios that have been discussed here this morning, when you integrate over the course of the century, you end up with captured and sequestered carbon that is denominated in hundreds of billions of tonnes. It doesn't take a rocket scientist to figure out that if you have a loss rate of only 1% and a stock of 200 billion tonnes, you are emitting 2 billion tonnes of carbon a year back from this reservoir. If at the end of the century, the total emission that the planet gets is on the order of 4 billion to 5 billion tonnes of carbon per year, you've got a problem.

Being able to tag the carbon and to guarantee that you know where it is, is going to be an extremely important issue. I think it comes back to some of the questions that people like Klaus were asking, and that is, the technologies in which controlling and monitoring become fairly easy are if CO_2 is taken off as a solid. There may be some value in pursuing an investigation down those lines, as well as removing the carbon as a gas.

Brian Flannery: Just a very quick couple of things to add. I think it's easy to try to make a list of things that might be useful, but I think more to the point is that you would probably have to be thinking in terms of creating a process with some scenarios or portfolios, some identification of issues, and their link. If you make progress in one area, if you overcome a barrier, it can suddenly open up a whole new chain of thought.

So I think part of it is not just that we need metallurgy or material science or reaction separations. We need all of this, it's true, but you need some systems context in which to think. This context certainly has to involve basic things like compressors and pipelines and central plants or distributed plants because then you see whether a breakthrough in this area actually makes a difference. If suddenly distributed generation of hydrogen makes more sense than centralized generation and then distribution, you may go off in totally new directions.

In terms of a national interest program in technologies that might come on line in 10, 20, 30 years, you need some context for it. Then, within this program, you have specific areas that you go after, that are maybe very deep fundamental science, or monitoring equipment, or control systems, or scenario planning. But you need a process, and it has to look at the overall situation, to be flexible, and to have science recycled into the agency planning process. Frankly, I think this has been a problem in our natural science research on climate change. The money is being spent, but the science is not being recycled in terms of what are we learning to adjust the research.

So I think we need a process-oriented approach in which there are some scenarios or portfolios but the planning should contain a recycle step to account for learning. The learning should include especially scientific and technical input, not just political input.

Tom Brownscombe, Shell: I just want to make a quick comment on the cost issue that was raised before. Our thinking is that the total cost for the transportation, sequestration, infrastructure—everything—is probably around $10 a ton of CO_2, similar to the cost that people are willing to pay for CO_2 for enhanced oil recovery. I also think, as someone has mentioned before, that the majority of the cost has to do with the capture and separation of the CO_2, which would be the prime thing to lower through research.

John Stringer: Just one small point about this. You can never just compare these things on the basis of the price that somebody is prepared to pay it, because the amount you are involved with is far more than the market you have to go to. When we started out the sulfur separation, we thought in terms of the gypsum product, we looked at the price that people were paying for gypsum, and we calculated the costs on that basis. We saturated the gypsum market so fast, you can't believe. It is nice that everyone builds these little cubicles in their offices, and if the cubicles ever get down to the space just occupied by your body, we can probably get rid of quite a bit more gypsum and justify oil and sulfur. Yet the fact is that you can't actually justify a separation activity on the basis of the current market.

David Thomas: Actually, I didn't get it in before. When we were talking about the process and the plan and the roadmapping for research in the area, I'm sure many of you are aware that the Department of Energy, Fossil Energy, and Office of Science put together what they call a carbon sequestration roadmap. I'm paraphrasing the title badly, but there is a section in there that deals with chemistry and chemical processes, and a lot of thought has been put in that. I just wanted to get it into the record that we make sure we look at what has already been done before we reinvent it, because there has been a great deal of effort already invested.

Bill Millman, Department of Energy: I come from the basic research community. One of the questions that you always get when you deal with Congress is the introduction of this basic research into technology. One of the things that has been going around in the carbon area is the fact that if you look into the future, you see developing countries being a source of major new inputs into the carbon arena.

One of the things that has been going on in the government is talk about using the cell phone, where if you don't have an infrastructure in a country, you can introduce brand new technology without suffering from any of the cost penalties that you would in a developed country. I would like to get your input onto how that would work in sort of a heavier industry.

Brian Flannery: I could make a quick comment. I don't think I can provide you with an answer to the way you framed the question, but it's true that in developing countries, in industries that are building grassroots facilities, such as the petroleum industry in many cases, you don't go back to your designers and say, let's build what we built 40 years ago in Louisiana. The materials have changed; the control systems have changed. You build a much, much more efficient, better-thought-through plant at this time. Now, that is not true in all industries, and it's not true in every case, but it's an enormous opportunity.

On the other hand, introducing brand new technologies in developing countries can add to the risk. You don't have the enabling infrastructure. You may have difficulties moving people and materials in and out easily. It is very risky, and if you are asking developing countries to participate in this, it is not clear that it's the best strategy at all. A good strategy for a developing country is to introduce good, proven, best-available technology, which usually makes an enormous improvement in things like local emissions and air quality. Yet asking a developing country to be the site perhaps of the introduction of brand new technology and its enabling infrastructure could potentially be a very risky enterprise.

We are also thinking about the 100-year problem. In many of the models, the developing countries become a lot more developed over the next 30 years, and their capacity to introduce new technologies could be much better. I'm increasingly convinced, however, that it's not so much an economy problem as a capacity-building problem in developing countries, where by capacity, you mean a whole range of issues.

These issues include the ability to make good decisions, to plan, to have legal structures, and to have an appropriate tax and regulatory framework—one that governments should be addressing on a priority basis for all kinds of reasons besides climate change.

Bill Millman: I would like to comment, and maybe John could address this. One of the areas in which there has been a significant amount of study is the nuclear power industry. You can't build a new nuclear power plant in this country. A lot of new technology has been developed since the last nuclear power plant was built, maybe 30 years ago. Building nuclear plants in Third World countries, on the other hand, where they have no electric industry, or just a very incipient one, provides some opportunities, and in fact, there has been talk about government subsidies for that. Could you comment on this?

John Stringer: Given the government subsidy, low-cost capital for building a power plant is the cheapest way of generating electricity. There is no doubt about this. The old comment about too cheap to meter (which the older among you may remember) didn't actually pan out, but it didn't pan out because of the number of additional costs that were added as a result of government actions at that time.

If you can subsidize the capital costs, it works out very cheaply. However, the point you made—that isn't all there is—is very important. If you go into a place, you need the infrastructure and, in particular, the infrastructure that relates to reliability, and the business of keeping a plant running and making sure it operates in a reliable and safe fashion requires an educational infrastructure that often doesn't exist.

I agree entirely with your point, and I think within the next 10 or 20 years the infrastructure problem will go away, but in the immediate future, the business of getting stuff maintained is very difficult—even things as trivial as automobiles. There are some parts of the world where you can buy an automobile, but you can't get it fixed, no matter what, and it would be a problem if one had that issue with a nuclear plant.

Antonio Lau, BP: We discussed zero emissions. We discussed spending almost unlimited amounts to save the future of the planet. We discussed the solutions. Do you know how much money is spent on analyzing the problem of climate change? One can track the CO_2 concentrations, temperature changes, and so forth.

One can also analyze radiated heat transfer and things like that. Perhaps in the early stage we may want to put an emphasis on that side of research, and then once we verify that a problem exists, we should go forward and find a solution.

Brian Flannery: I'm willing to make a very quick stab at your question. I think everyone is convinced that the buildup of greenhouse gases traps extra infrared radiation in the atmosphere, and this could lead to a problem down the road. There are few people who aren't convinced, at least in terms of the radiative transfer issues. The deep scientific questions in climate change concern the feedbacks that may occur, and the future of forcing from human actions. Once heat is trapped, climate processes, such as clouds, ocean circulation, hydrology and moist convection can change in ways that science cannot yet predict.

How big the problem will be, how soon it will be here, and with what positive and negative consequences, remain fairly uncertain, but I think in terms of managing a risk. The political system is convinced that the scientific evidence is there and that there is a risk that needs to be managed.

The question we are all faced with, and have been for a long time is, How? How much effort do you put in, with what scope, and what initial emphasis? The Kyoto Protocol negotiation is certainly a clear sign of this, though it does look as though it is running into some difficult waters, because it may have tried to do too much too soon at too high a cost, and it doesn't address the long-term issues.

So now the question is, from a perspective of scientific research, What can the research community add to developing improved possibility of solutions? I would also add, but I don't think this is the focus of your meeting, what can the chemistry community add to our ability to understand the climate issue?

There is a great deal of money being spent on research. I think the United States has been spending in the neighborhood of $2 billion a year for the last several years. I do think that there could have been a better framework in which the research money was spent in terms of delivering answers to the societal questions, as opposed to the scientific inquiry questions or the political questions that are being asked.

There could be a better framework for the research to proceed. However, the world is spending a few billion dollars a year on research on climate change, and certainly the chemistry community has a lot it can contribute to this too on a fundamental level of gauging what the real nature of the issues is, what the nature of natural climate variability is, and related questions.

John Stringer: I can add something to that. First, I agree with all of this. There is a significant amount of research being done that hasn't been discussed here, but this doesn't mean it isn't being done. My institute does climate research for example, and we have supported climate research, which includes looking at some of the models suggesting that the warming over the last 100 years, or whatever number of years you think we have been warming for, is not necessarily causally related to the increase in CO_2 that occurred over the same period. In other models, these two things happened, but they are not necessarily related. That is the basis of the disagreement argument. When I was very young, we had gone through a period of about 10 years when temperatures had been falling progressively over much of the world. At that time, we were looking at the risk of the next ice age. I found it very entertaining to remember back to those days and realize that now we are talking about the exact reverse of that.

So I'm sensitive to the point that you make—that maybe the case isn't proven. However, my point isn't really that. My point is related to the business of risk scenarios. When we first ran into problems in the nuclear industry, we tried to do careful calculations of risk, and risk-based planning of course is well known. The Moon mission, for example, was based on asking the astronauts what risk they would accept, and they said they would accept the same risk as getting killed by a car in Houston. I don't think they put it that way now, but that was the number they chose at the time, and that was the number that was used as the basis for risk calculations in the whole Moon program.

Now we tried to do a similar calculation for the risks involved with nuclear energy, and we found that the risks were very small. These calculations have been borne out by the performance over the subsequent 20 years of the nuclear industry, where the accident rate has been extremely low. Nevertheless, it became clear that the public didn't care that we had calculated that the risks were very small. It became a matter of what is called risk perception. Consequently, what we did was to shift lots of our research in this area to risk perception.

So at the moment, it doesn't matter whether or not our industry point of view is that there is a risk of global warming. If the perception of the electorate, the people, is that there is global warming, then legislation will follow because that's the way society works, and we better be ready to react to that. So I always think in terms of what the legislation will look like, and we're negotiating with people to try and get legislation to deal with things that we think we stand a chance of being done.

So that's the problem. It doesn't mean that we should stop climate research based on alternate scenarios, because at some point we may change people's perception. Yet that's the way it goes at the moment: risk perception is more important than risk.

Richard Wool, University of Delaware: I'm looking for marching orders. I know we are trying to set the research agenda, but I'm wondering what each of you thinks are the critical issues or strategies? If there was one thing that each of you would advise somebody to do, what is it?

Richard Alkire: One thing for the academic research community?

Richard Wool: The thing the academic research community could do that would have the greatest impact on the carbon management scenario.

David Thomas: This is not flippant. I think we need to be working on fusion a heck of a lot more than we are. I wish we would spend a great deal more effort in that area, because I think it has the potential of going beyond. I would argue that, although we will not run out of hydrocarbons in the foreseeable future, or at least the next several hundred years, we will see a hydrocarbon age, much as we saw a wood age, and so on. Also, I would argue that it may be 200, it may be 300, it may be 250, it may be 600 years, but there will be an end to the hydrocarbon age, and my question is, What is beyond the hydrocarbon age?

James Edmonds: Research is needed on the full range of sequestration technologies, going all the way from pulling it off as a solid, to disposing of it in the form of a gas, to nature reservoirs including soils and forests.

Brian Flannery: I'm going to offer you two suggestions. The first is be creative and invent something new, because none of the current approaches look like they are going to work in the near term. The second is very serious. Convince people that science and technology has been a positive force in their lives, and has something to offer. This can be a positive force in the debate, rather than being used to pick and choose from science and technology to build horror stories to convince people that there is no way out.

John Stringer: Electricity is a wonderful thing but is difficult to store. I would like to see the whole business of where you store things in the overall cycle addressed. As we move toward a hydrogen economy, this is going to become particularly important. Hydrogen is extremely difficult to store, electrons are very difficult to store, and some further work in these areas would transform the economies of some of the things that we can't do yet.

5

Carbon Dioxide as a Feedstock

Carol Creutz and Etsuko Fujita
Brookhaven National Laboratory

This chapter is an overview on the subject of carbon dioxide as a starting material for organic syntheses of potential commercial interest and the utilization of carbon dioxide as a substrate for fuel production. It draws extensively on literature sources,[1-3] particularly the report of a 1999 workshop on the subject of catalysis in CO_2 utilization,[1] but with emphasis on systems of most interest to us.

Atmospheric carbon dioxide is an abundant (750 billion tons of carbon in the atmosphere) but dilute source of carbon (only 0.036% by volume),[3] so technologies for utilization at the production source are crucial for both sequestration and utilization. Sequestration—such as pumping CO_2 into the seas or the earth—is beyond the scope of this chapter, except where it overlaps utilization—for example, in converting CO_2 to polymers. Yet sequestration dominates current thinking on short term solutions to global warming, as should be clear from reports of this and other workshops.[4,5] The net anthropogenic increase of 13,000 million tons of carbon dioxide estimated to be added to the atmosphere annually at present can be compared to the 110 million tons of CO_2 used to produce chemicals, chiefly urea (75 million tons of CO_2), salicylic acid, cyclic carbonates, and polycarbonates.[1] Increased utilization of CO_2 as a starting material is, however, highly desirable, because it is an inexpensive, nontoxic starting material. There are ongoing efforts to replace phosgene as a starting material.[6] Creation of new materials and markets for them will increase this utilization, producing an increasingly positive, albeit relatively small, impact on global CO_2 levels. The other uses of interest are utilization as a solvent and for fuel production, and these are discussed in turn.

PRINCIPAL CURRENT USES OF CARBON DIOXIDE

Urea synthesis is currently the largest use of carbon dioxide in organic synthesis. Urea, $C(O)(NH_2)_2$, is the most important nitrogen fertilizer in the world. Urea is also an intermediate in organic syntheses such as the production of melamine and urea resins, which are used as adhesives and bonding agents. Carbon dioxide is also used to produce salicylic acid, which is found in pharmaceuticals, and cyclic organic carbonates, high melting, but extremely high boiling solvents for natural and synthetic polymers such as lignin, cellulose, nylon, and polyvinyl chloride (Figure 5.1). The latter are used extensively in

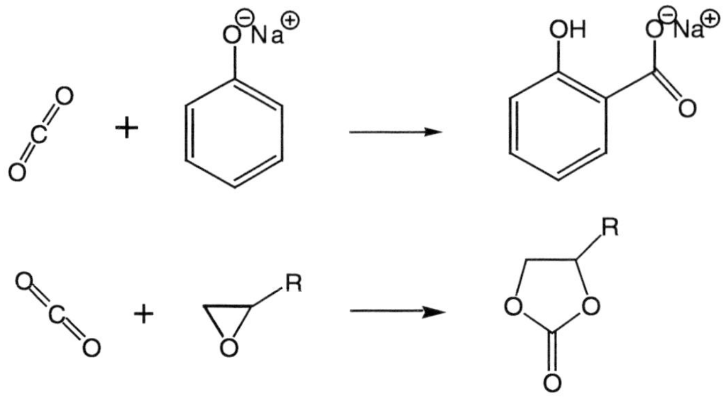

FIGURE 5.1 Examples of reactions involving carbon dioxide and leading to intermediates in industrial synthesis. Top: The reaction of carbon dioxide with sodium phenolate to produce sodium salicylate. Bottom: The reaction of carbon dioxide with an epoxide to make a cyclic organic carbonate.

the production of polyacrylic fibers and paints. Ethylene and propylene carbonates have many uses in chemical synthesis—among them reactions with ammonia and amines to form carbamates and subsequent reactions with diamines to yield di(hydroxyethyl) carbamates, which can react further with urea to form polyurethanes.

Figure 5.2 provides a broad summary of current and projected utilization of carbon dioxide. Reactions that use CO_2 to produce organic chemicals or intermediates for the chemical industry are summarized from 6 to 11:00 o'clock on the diagram. The first examples, such as salicylic acid, are in current practice. Also included in Figure 5.2 are reactions that hold promise for extensive utilization in the future. Many of these involve insertion of carbon dioxide into Y—X bonds, often the C—H bond. The products of interest include esters, carbamic esters, salicylic acid, and cyclic carbonates. These reactions

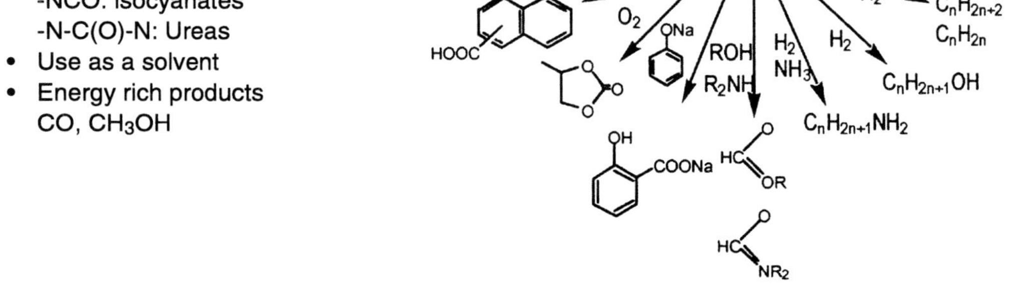

- Intermediate of fine chemicals for the chemical industry
  -C(O)O-: Acids, esters, lactones
  -O-C(O)O-: Carbonates
  -NC(O)OR-: Carbamic esters
  -NCO: isocyanates
  -N-C(O)-N: Ureas
- Use as a solvent
- Energy rich products
  CO, CH_3OH

FIGURE 5.2 Utilization of CO_2 in synthetic chemistry. SOURCE: Aresta (1998).[6]

commonly involve formal insertion of carbon dioxide into an X—H bond, as into the N—H bond in urea formation. Novel insertions under active investigation involve the incorporation of CO_2 into polymers—polycarbonates, polypyrones, lactone intermediates, and polyurethanes. Of particular interest is the incorporation of carbon dioxide into polymers, an active area of research and one that is very promising for future applications.[1,7] However, the impact of new materials and processes in this area will ultimately depend on market forces, a factor than can be frustrating to the researchers. Other reactions shown in Figure 5.2 are, from 2 to 6 o'clock, hydrogenation reactions and, from 12 to 2 o'clock, hydrogenations accomplished by electrons and protons—both directed toward fuel formation.

Carbon Dioxide as a Solvent

Supercritical carbon dioxide is a hydrophobic solvent that can replace organic solvents in a number of applications. Its critical temperature is 31 °C, and it has very low viscosity. When carbon dioxide is substituted for an organic solvent, solvent costs and emission of toxic organics can be reduced. Furthermore, separation of the products and catalyst can be controlled easily by changing the carbon dioxide pressure. Currently, supercritical carbon dioxide is used in caffeine extraction, dry cleaning, and parts degreasing. These processes can involve high-capacity plants of more than 22.5×10^6 kg per year in the case of decaffeination processes.[8] Potential future or developing applications include utilization in food and pharmaceutical processing to defray future liability costs, production of pharmaceutical nanoparticles for injection, polymerizations,[9] emulsion polymerization of water-soluble monomers, enhanced oil recovery, and homogeneous[10] and phase-separable catalysis, including that based on ionic liquid solvents.[11]

REACTIVITY OF CARBON DIOXIDE

Carbon dioxide is a linear molecule in which the oxygen atoms are weak Lewis (and Brønsted) bases and the carbon is electrophilic. Reactions of carbon dioxide are dominated by nucleophilic attacks at the carbon, which result in bending of the O—C—O angle to about 120°. Figure 5.3 illustrates four very different nucleophilic reactions: hydroxide attack on CO_2 to form bicarbonate; the initial addition of ammonia to CO_2, which ultimately produces urea; the binding of CO_2 to a macrocyclic cobalt(I) complex, which catalyzes CO_2 reduction;[12,13] and the addition of an electron to CO_2 to yield the carbon dioxide radical ion. The first three also exemplify the reactivity of CO_2 with respect to nucleophilic attack on the carbon.

Thermodynamic Barriers to CO_2 Utilization

Carbon dioxide is a very stable molecule, and accordingly, energy must generally be supplied to drive the desired transformation. High temperatures, extremely reactive reagents, electricity, or the energy from photons may be exploited to carry out carbon dioxide reactions.

Figure 5.4 depicts reactions in which the energy source is renewable or nuclear. ("Renewable" refers to solar electric, wind hydroelectric, geothermal, solar thermal, and biomass-based energy sources.) The reaction $CH_4 + CO_2 = 2\ CO + 2\ H_2$ is called the carbon dioxide reforming of methane. This reaction, if combined with metal production in situ, using solar furnaces to achieve the high temperatures needed (e.g., 1200 °C), could be used to significantly mitigate CO_2 produced in cement, lime, and metal (iron, aluminum) production, which amounts to about 10% of total CO_2 released annually.[3] For electrochemical reduction of CO_2 to methane, energy may be derived from a solar cell or

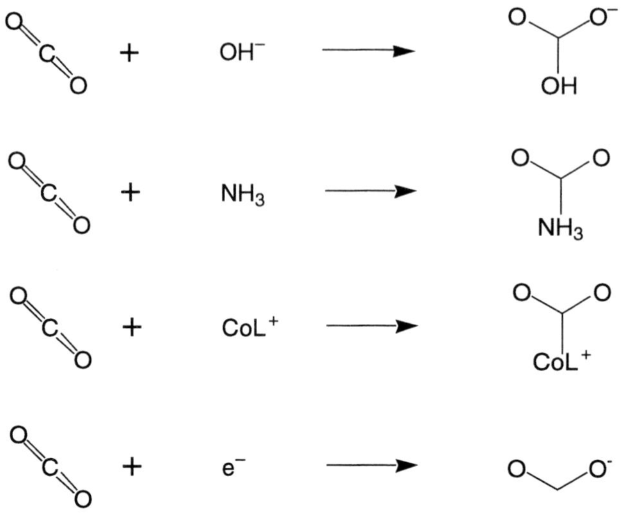

FIGURE 5.3 Reactions of carbon dioxide are dominated by nucleophilic attacks at the carbon atom.

nuclear power. Reduction may also be accomplished photochemically by utilizing a dye to absorb visible light, since CO_2 itself does not. Interestingly, vacuum ultraviolet (VUV) irradiation of carbon dioxide yields oxygen and carbon monoxide.

Conversion of Carbon Dioxide to Fuels

Direct Hydrogenation

With abundant renewable energy sources, carbon dioxide can be converted to fuels by reduction to methanol or methane. The high energy density of carbon-based fuels and their availability as either gases, liquids, or solids are important reasons for the dominant position of fossil fuels in the current marketplace. Because the value of a fuel is based on its energy content and its ease of transport and

Approach	Energy Source
• High temperatures $CH_4 + CO_2 = 2\,CO + 2H_2$	• Solar furnace 1200 °C
• Electrochemical reaction $CO_2 + 8e^- + 8\,H^+ = CH_4 + 2H_2O$	• Solar cell; nuclear power
• Photochemical reaction $CO_2 + 2e^- + 2H^+ = CO + H_2O$ $CO_2 = CO + {}^1/_2\,O_2$	• Light from sun • VUV light

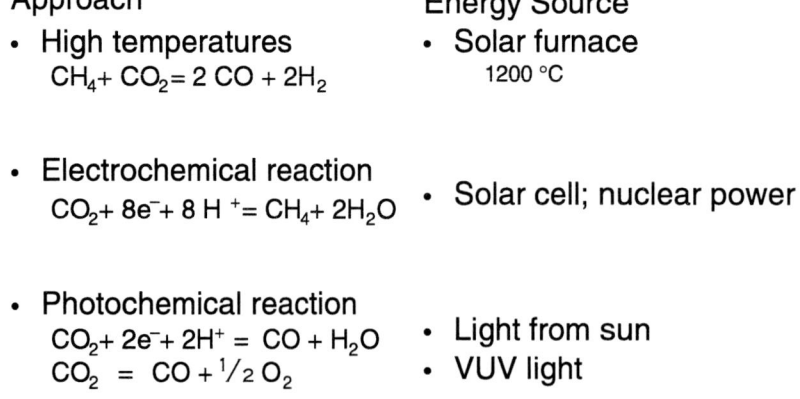

FIGURE 5.4 Overcoming the thermodynamic barriers to CO_2 utilization.

storage, methane is less desirable than methanol because of its low fuel density and high cost of transport. Today, carbon dioxide is a by-product of fuel use, not a feedstock for fuel production. Conversion of CO_2 to fuels using renewable or nuclear power produces no net emission of carbon dioxide (excluding CO_2 produced by energy consumption in the reduction process), and it would complement the renewable production of fuels from biomass, which is likely to be insufficient to meet future world demands. Catalysis can play an important role in this area. The objective of such fuel production is to develop strategies for reduction of CO_2 that can be adapted for use at different sources and yield fuel products widely utilizable with current and future technologies.

Hydrogenation of carbon dioxide to methanol is slightly exergonic, and to methane to a greater extent, because of the favorable thermodynamics of water formation (see Figure 5.5).

Catalysis of hydrogenations leading to N, N-dimethylformamide (DMF), formate, and hydrocarbons is being addressed successfully.[10,14] Reduction to carbon monoxide is also useful when the CO hydrogen mixtures can be used to augment feeds in industrial processes such as ethylene and methanol production. Methanol, lower hydrocarbons (methane, ethane, ethylene, etc.), CO, and formic acid (HCOOH) have been prepared from CO_2 and H_2 using several different metal and metal oxide catalysts at elevated temperature and pressure.[15] Selectivity and catalytic activity depend on the catalyst used (i.e., metal or metal oxide), its size, additives, support, temperature, CO_2 to H_2 ratio, and pressure. Copper on zinc oxide (ZnO) seems to be the most active catalyst for methanol production.[15,16] A small-scale test plant with a production capacity of 50 kg per day of methanol was constructed at the Research Institute of Innovation Technology of the Earth to examine the performance of a Cu/ZnO-based multicomponent catalyst (Cu/ZnO/ZrO_2/Al_2O_3/Ga_2O_3) under practical conditions.[17] Selectivity for methanol production was found to be very high, and direct methanol production from CO_2 may be commercially feasible with an inexpensive source of hydrogen.[17] Use of hydrogen for this reduction chemistry is not, however, economically attractive at present because of its cost; inexpensive production of hydrogen by solar or nuclear power sources could radically alter this scenario.

Indirect Hydrogenation

Hydrogen may be replaced by electrons and protons, available, for example, in electrochemical reduction in aqueous media (Figure 5.6). One-electron reduction of carbon dioxide to the radical anion CO_2^- presents thermodynamic and kinetic barriers. In aqueous solution, the reduction potential is -1.9 V versus the standard hydrogen electrode (SHE).[18] The barrier to outer-sphere electron transfer for the couple is large because of the very different geometries of the linear, neutral carbon dioxide and the bent radical anion.[19] Thus, direct (uncatalyzed) electroreduction requires a significant overvoltage. As shown in Figure 5.5, the thermodynamic barriers are reduced by protonating the reduction product.[13,20] However, because of the near equivalence of the hydrogen potential to that for the proton-assisted hydrogenation of CO_2, reduction of H^+ or H_2O to H_2 may also occur, depending on the system.

Hydrogenation	ΔG^0 (kcal/mol 298 K)
$CO_2 + 3 H_2 = CH_3OH + H_2O$	-4.1
$CO_2 + 4 H_2 = CH_4 + 2 H_2O$	-31.3

(Standard states: gases at 1 atm., water liq., methanol 1M)

FIGURE 5.5 Hydrogenation of carbon dioxide to methanol and methane.

One-electron reduction
$$CO_2 + e^- = CO_2^- \quad E = -1.9 \text{ V}$$ 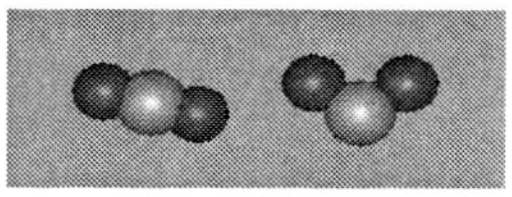
Two-electron reduction
$$CO_2 + 2e^- = CO_2^{2-} \quad -1.55 \text{ V}$$
Proton-assisted pH 7 vs. SHE 25°C
$$CO_2 + H^+ + 2e^- = HCO_2^- \quad -0.49 \text{ V}$$
$$CO_2 + H^+ + 2e^- = CO + OH^- \quad -0.53 \text{ V}$$
$$CO_2 + 6H^+ + 6e^- = CH_3OH + H_2O \quad -0.38 \text{ V}$$
versus: $2H^+ + 2e^- = H_2 \quad -0.41 \text{ V}$

Standard states: For aqueous solutions, gases 1 atm, others 1M

FIGURE 5.6 Thermodynamic and kinetic aspects of CO_2 reduction. SOURCE: Sutin et al. (1997),[20] Fujita (1999).[13]

Electrochemical Reduction

As noted earlier, direct electroreduction is achieved at high overvoltage. An unreactive metal or carbon electrode produces carbon dioxide radical anion, which may undergo dimerization to oxalate or disproportionation to carbon monoxide and carbonate.[21] By contrast, non-innocent metals, through active sites on their surfaces, can direct CO_2 reduction to hydrogenated products at a much lower applied voltage because of the high efficiency of the heterogeneous catalysis. Particularly noteworthy is the work of Hori,[22] which showed that copper produces high yields of methane from aqueous bicarbonate at 0 °C and high yields of ethylene at 45 °C. In these systems, the metal serves a dual role, both delivering electrons and stabilizing the reduced fragments. In the case of copper, the reduction is believed to involve the sequence shown in Figure 5.7. The metals ruthenium, cadmium, mercury, indium, tin, and lead yield formate; gold, silver, and zinc yield CO; while aluminum, gallium, platinum, iron, nickel, and titanium exhibit little activity.[3] Other important areas of electrochemical reduction are homogeneous catalysis, surface modified by a "molecular" catalyst, and photoelectrochemical systems.

Homogeneous Catalysis of Carbon Dioxide Reduction

Homogeneous catalysts may fulfill the role of the surface metal catalytic sites in the above systems (for example, copper). Homogeneous catalysis is important in electrochemical reductions systems, as well as photochemical systems. Indeed the two approaches share many features, as discussed later.

Catalysts lower the overpotential for CO_2 reduction by undergoing reduction at a potential (E_{cat}) less negative than that for direct CO_2 reduction, binding CO_2, or undergoing a second reduction at (or more positive than) E_{cat}. For homogeneous catalysis,[21] two-electron reduction to give CO, HCO_2^-, or $C_2O_4^{2-}$ is most frequently observed, and added proton sources may be required.

$$CO_2 \rightarrow CO \rightarrow CuHCO \rightarrow Cu=CH_2 \rightarrow CH_4$$

FIGURE 5.7 Sequence of the electrochemical reduction of CO_2 with copper to form methane.

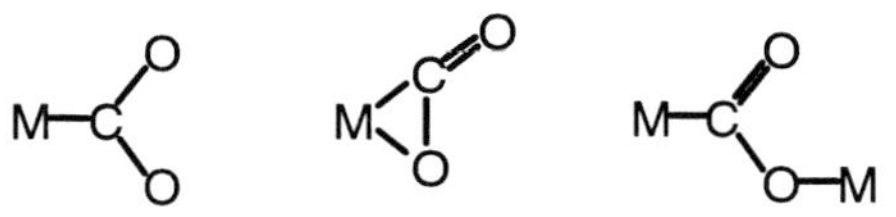

FIGURE 5.8 Modes of binding carbon dioxide to transition metal centers.

Transition Metal Catalysis

Binding of carbon dioxide to transition metal centers has been reviewed.[23,24] Among the several modes in which the metal may bind the CO_2 are the three shown in Figure 5.8, end-on, C η^1; bridging, C, O η^2; and bimetallic motifs in which one metal binds C η^1 and another metal binds O η^1.

Electrochemical Systems

Electrocatalysis by transition metal complexes is elegantly illustrated by the work of DuBois and colleagues.[25] In the absence of CO_2, the palladium (II) complex undergoes two-electron reduction, but when CO_2 is present, the one-electron reduction product binds CO_2 (DMF as solvent) (Figure 5.9). With added acid, carbon monoxide is produced catalytically from carbon dioxide. This chemistry likely involves the protonated carbon dioxide adduct (hydroxycarbonyl complex) shown in Figures 5.9 and 5.10.[26] Catalytic turnover numbers greater than 100 have been reported for this and related compounds. Some hydrogen is produced in parallel, evidently via a hydride complex.

Other systems for electrochemical CO_2 reduction utilize transition metal complexes of nitrogen-containing (nickel and cobalt) macrocycles (including porphyrins and phthalocyanines) and (ruthenium, cobalt, and rhenium) complexes of 2,2'-bipyridine.[27]

Photochemical Systems[28]

Photochemical reduction systems (Figure 5.11) require efficient light harvesting, usually by a so-called dye or sensitizer, and efficient charge separation and energy utilization. Transition metal complexes, particularly tris(2,2'-bipyridine)ruthenium(II), serve as sensitizers. The overall reaction carried out must be a useful one. That is, in addition to carbon dioxide reduction, the complementary oxidation process (which provides the electrons) should be a desirable one. Both reduction and oxidation processes generally require catalysis. For carbon dioxide reduction, a number of the catalysts used in electrochemical systems are also effective in photochemical systems, as outlined below.

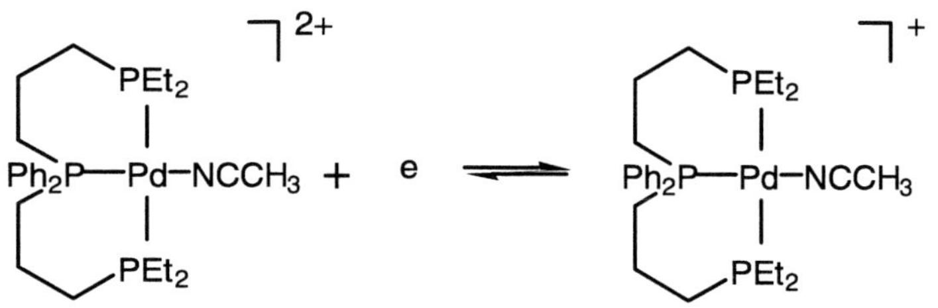

FIGURE 5.9 Reduction of palladium (II) to palladium (I) in the presence of CO_2.

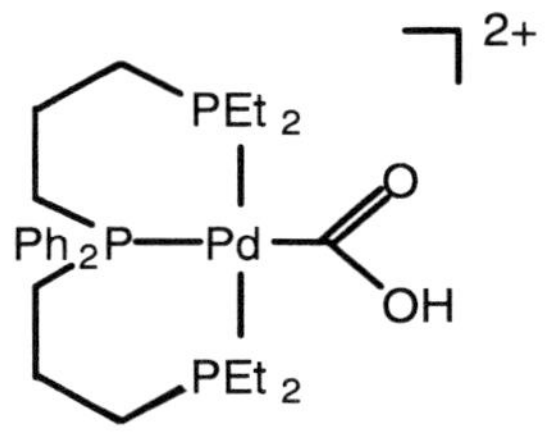

FIGURE 5.10 The protonated carbon dioxide adduct involved in reduction chemistry.

Comparison of Electrochemical and Photochemical Systems

When the catalytic reduction of carbon dioxide is truly homogeneous (occurs in the solution), electrochemical and photochemical systems may have much in common. The means of electron delivery differs, of course, with photoinduced electron transfer processes serving the role of the electrode in the photochemical system. Many of the catalyst systems studied so far—cobalt and nickel macrocycle systems, for example—work in both electrochemical and photochemical systems. In both approaches, the ultimate source of these electrons is an issue (Figure 5.12). Sacrificial reagents (generally organic compounds that become oxidized) are commonly used, and one of the challenges is to replace these reactions with processes that are less costly and wasteful. For aqueous systems, it would be highly desirable to use the water oxidation half-reaction, that is,

$$H_2O = 2\ e^- + 1/2\ O_2 + 2\ H^+$$

for this purpose, so that the overall reaction would be

$$CO_2 + 2\ H_2O \rightarrow CH_3OH + 3/2\ O_2.$$

The challenge remains the effective development and deployment of water oxidation catalysts.

At present, electrochemical reduction of CO_2 yields carbon monoxide, formate, methane, and so forth, with good current efficiencies and, in photochemical systems, quantum yields of carbon monoxide (or formate or both) up to 40%.

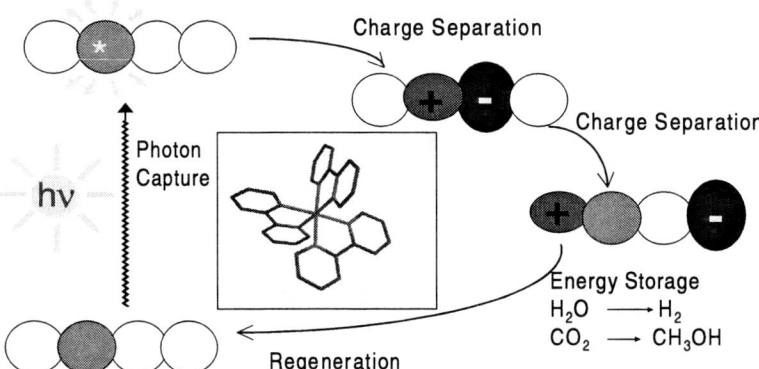

FIGURE 5.11 Photoinduced charge transfer with transition metal complexes.

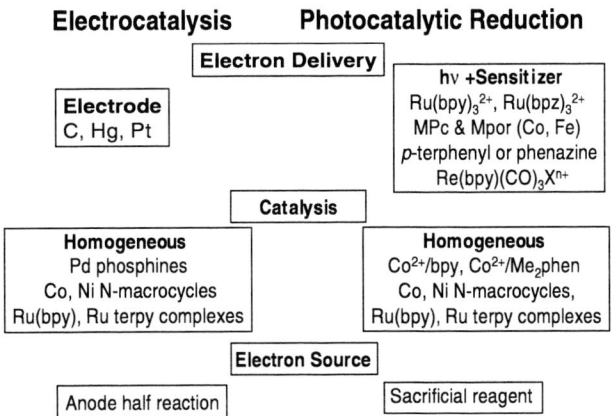

FIGURE 5.12 Comparison of photochemical and electrochemical systems for reducing CO_2.

OPPORTUNITIES

There are many areas in which ongoing and future research can lead to new modes of carbon dioxide utilization. These include the following:

- Utilization of CO_2 in new polymers
- Development and understanding of both homo- and heterogeneous catalysts for the following:
  1. Polymerization, hydrogenation, electrochemical and photochemical processes
  2. Utilization of soluble and surface-anchored nanoparticles of metal and semiconductor clusters
  3. Reactions or catalysis in supercritical CO_2
- Electrochemical and photochemical electron sources in the presence of proton sources can avoid use of expensive H_2, but both need:
  1. Faster catalytic processes and more stable catalytic systems
  2. Development of useful second half reaction (i.e., elimination of sacrificial reagent or useful anode reaction).

ACKNOWLEDGMENT

We thank D.L. DuBois for helpful comments. This research was carried out at Brookhaven National Laboratory under contract DE-AC02-98CH10886 with the U.S. Department of Energy and was supported by its Division of Chemical Sciences, Office of Basic Energy Sciences.

REFERENCES

1. Aresta, M., J.N. Armor, M.A. Barteau, E.J. Beckman, A.T. Bell, J.E. Bercaw, C. Creutz, E. Dinjus, D.A. Dixon, K. Domen, D.L. Dubois, J. Eckert, E. Fujita, D.H Gibson, W.A. Goddard, D.Goodman, J. Keller, G.J. Kubas, H.H. Kung, E. Lyons, L.E. Manzer, T.J. Marks, K. Morokuma, K.M. Nicholas, R. Periana, L. Que, J. Rostrup-Nielson, W.M.H. Sachtler, L.D. Schmidt, A. Sen, G.A. Somorjai, P.C. Stair, B.R. Stults, and W. Tumas. *Chem. Rev.* 2001, in press.
2. Behr, A. 1988. *Carbon Dioxide Activation by Metal Complexes.* VCH: Cambridge.
3. Halmann, M. M., and M. Steinberg. 1999. *Greenhouse Gas Carbon Dioxide Mitigation.* CRC Press: Boca Raton, Fla.
4. Benson, S., W. Chandler, J. Edmonds, M. Levine, L. Bates, H. Chum, J. Dooley, D. Grether, J. Houghton, J. Logan, G. Wiltsee, and L. Wright. 1998. *Carbon Management: Assessment of Fundamental Research Needs.* U.S. Department of Energy: Washington, D.C.

5. Reichle, D., J. Houghton, B. Kane, J. Ekmann, S. Benson, J. Clarke, R. Dahlman, G. Hendrey, H. Herzog, J. Hunter-Cevera, G. Jacobs, R. Judkins, J. Ogden, A. Palmisano, R. Socolow, J. Stringer, T. Surles, A. Wolsky, N. Woodward, and M. York. 1999. *Carbon Sequestration Research and Development*. U. S. Department of Energy: Washington, D.C.
6. Aresta, M. 1998. *Advances in Chemical Conversions for Mitigating Carbon Dioxide; Studies in Surface Science and Catalysis 114*, 65-76.
7. Darensbourg, D.J., and M.W. Holtcamp. 1996. *Coord. Chem. Rev. 153*, 155-174.
8. Ghenciu, E.G., and E.J. Beckman. 1998. "Carbon Dioxide" www.AccessScience.com.
9. Kendall, J.L., D.A. Canelas, J.L. Young, and J.M. DeSimone. 1999. *Chem. Rev. 99*, 543-563.
10. Jessop, P.G., T. Ikariya, and R. Noyori. 1999. *Chem. Rev. 99*, 475-493.
11. Blanchard, L.A., D. Hancu, E.J. Beckman, and J.F. Brennecke. 1999. *Nature 399*, 28-29.
12. Fujita, E.; C. Creutz, N. Sutin, and D.J. Szalda. *J. Am. Chem. Soc. 113*, 343-353, 1991.
13. Fujita, E. 1999. *Coord. Chem. Rev. 185–186*, 373–384.
14. Jessop, P.G., T. Ikariya, and R. Noyori. 1995. *Chem. Rev. 99*, 259-271.
15. Arakawa, H. 1998. *Advances in Chemical Conversions for Mitigating Carbon Dioxide; Studies in Surface Science and Catalysis 114*, 19-30.
16. Kieffer, R., and L. Udron. 1998. *Advances in Chemical Conversions for Mitigating Carbon Dioxide; Studies in Surface Science and Catalysis 114*, 87-96.
17. Inui, T., M. Anpo, K. Izui, S. Yanagida, and Y. Yamaguchi. 1998. *Advances in Chemical Conversions for Mitigating Carbon Dioxide*, Delmon, B., and J. T. Yates., Eds. Elsevier: Amsterdam.
18. Schwarz, H.A., and R.W. Dodson. 1989. *J. Phys. Chem. 93*, 409-414.
19. Schwarz, H.A., C. Creutz, and N. Sutin. 1985. *Inorg. Chem. 24*, 433-439.
20. Sutin, N., C. Creutz, and E. Fujita. 1997. *Commts. Inorg. Chem. 19*, 67-92.
21. Keene, F.R., and B.P. Sullivan. 1993. Pp. 118-140 in *Mechanisms of the Electrochemical Reduction of Carbon Dioxide Catalyzed by Transition metal Complexes*, Sullivan, B.P., K. Krist, and H.E. Guard, Eds.; Elsevier: New York.
22. Hori, Y., A. Murata, and R. Takahashi. 1989. *J. Chem. Soc., Faraday Trans. I 85*, 2309-2326.
23. Creutz, C. 1993. *Carbon Dioxide Binding to Transition-Metal Center*. Pp. 19-67 in Electrochemical and Electrocatalytic Reactions of Carbon Dioxide, B.P. Anllivan, Amsterdam: Elsevier.
24. Gibson, D.H. 1996. *Chem. Rev.* 2063-2095.
25. Wander, S.A., A. Miedaner, B.C. Noll, R.M.Barkley, and D.L. DuBois. 1996. *Organometallics* 3360-3373.
26. DuBois, D.L., A. Miedaner, and R.C. Haltiwanger. 1991. *J. Am. Chem. Soc. 113*, 8753.
27. Collin, J. P., and J.P. Sauvage. 1989. *Coord. Chem. Rev. 93*, 245-268.
28. Fujita, E. and B.S. Brunschwig. 2001. Pp 88-126 in *Homogeneous Redox Catalysis of CO2 Fixation*, Balzani, V. Ed.; Wiley-VCH, Weinheim Vol IV.

DISCUSSION

Glenn Crosby, Washington State University: Carol, I have not been involved with the Department of Energy for several years in this kind of research, but what has happened to the level of funding for the utilization of, say, protons for promoting photocatalysis and photoelectron over the last few years?

Carol Creutz: I am going to defer to Bill Millman from Chemical Sciences.

Bill Millman, Department of Energy: Well, in flat-budget scenarios, it has essentially gone down approximately 1.5% per year over the last about six years. In real terms—in constant dollars—it is a significant percentage. If you look at staffing at the labs, it means about 25%. This is one of the effects of the constant budgets.

It is safe to say then that the effort in photocatalysis and photoelectric chemistry I observed and was involved in four or five years ago has not kept pace with inflation but has actually decreased significantly in absolute terms.

6

Advanced Engine and Fuel Systems Development for Minimizing Carbon Dioxide Generation

James A. Spearot
General Motors Corporation

It's a pleasure to have the opportunity to participate in this "roundtable" and to provide General Motors' perspectives on the issues and research opportunities related to the concept of carbon management.

The purpose of this chapter is not to discuss the science of global warming or to take part in a debate over predicted ambient temperature increases. General Motors' (GM's) position on these subjects has been presented before. Specifically, "We believe there is enough cause for concern to warrant responsible actions to reduce global greenhouse gas emissions." Since reductions in carbon dioxide emissions are an essential part of this objective, I would like to discuss some of the options available to us in developing advanced engine and fuel systems for minimizing CO_2 generation.

The objectives here are the following:

- Review potential changes in energy utilization during the next century.
- Review research goals and promising vehicle technologies for reduced CO_2 generation from light-duty vehicles.
- Identify future transportation fuel needs, particularly if vehicles are to produce no net increase in CO_2.
- Summarize probable engine and fuel systems for the twenty-first century.

I would like to begin by providing a historical perspective. The complex issues related to selecting an optimum engine and fuel system are not new.

"Fuel today is a subject that is engrossing the minds of many men with an investigative turn. It is a subject, too, that requires quite a bit of figuring, for it is a serious problem."

"There are a number who are working out new methods of producing gasoline, as well as inventing new liquid fuels. Each is absolutely sure his method will eventually solve the present problem, opening up to the motoring public a source of cheap fuel that will last for many, many years to come."

The author of these quotes was Charles F. Kettering, the founder of the Dayton Engineering Laboratory (DELCO) and of General Motors Research Laboratories, as published in the DELCO magazine in 1916.

Although these quotes were written early in the twentieth century, they provide a fairly good summary of where we are at the beginning of the twenty-first century. The technical issues related to operation of engine/fuel systems are different, the societal issues driving change are different, the sophistication of engine and fuel technology is much greater, but the search for better engine and fuel systems continues.

Today, Earth's population tops 6 billion people and more than 700 million cars and trucks are in use (Figure 6.1). This means that only 12% of us are realizing the benefits provided by engine-powered vehicles. No other transportation technology gives us the freedom to go wherever we want, whenever we want, with whomever we choose, carrying whatever we need. This connectivity is what drives the nearly universal aspiration people have for "auto-mobility."

To satisfy this demand in the future, we must realize sustainable auto-mobility. With the world's population predicted to reach 7 billion by 2015, the global car park will grow to nearly 850 million vehicles over the next 15 years if the ownership rate remains at 12%. While this represents significant growth in the demand for our products, as an industry we need to be thinking bigger than this. If the ownership rate increases just 3 points—to 15%—the car park in 2015 would exceed 1 billion vehicles.

We will not be able to take advantage of this market opportunity unless auto-mobility is truly sustainable. Energy forms that are renewable and vehicle technologies that have zero impact on the ambient environment will be required for society to continue to enjoy and expand the benefits of personal mobility.

One way to understand how the nature of supplied energy for both mobile and stationary sources might change during the next 100 years is to compare the primary energy forms used at both the beginning and the end of the twentieth century (Figure 6.2). Although the changes might not appear substantial at first glance, the transition in energy forms has been significant. Why did the shift occur? It occurred because human needs for greater amounts of practical energy and reduced environmental impact demanded the shift and technological advances enabled the changes. As a result, society is better off due to the transition that took place.

Where are we headed in the next 100 years? My version of Figure 6.3 was published in 1996 in *Daedalus*, the journal of the American Academy of Arts and Sciences. A figure from the paper suggests that the calculated hydrogen-to-carbon (H/C) ratio of total energy consumed in the world is increasing at an exponential rate. Although this assumption can surely be debated, I believe the important point of this figure is the trend in the data, not the mathematical relationship used to fit the trend. General Motors believes this trend will continue well into this century, culminating at the point of its implied conclusion—a hydrogen economy. Extrapolating the data in this plot 100 years into the future takes "fortune-

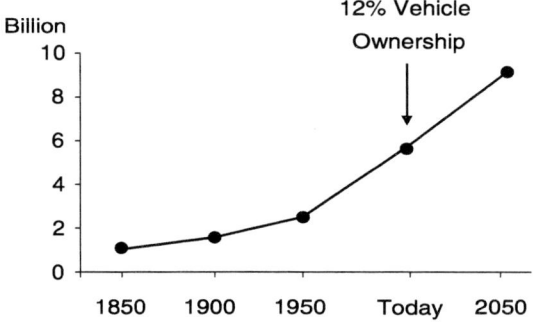

FIGURE 6.1 World population and vehicle ownership.

```
1900
    - Biomass (wood, peat)
    - Coal
    - Kerosene
    - Natural Gasoline
    - Fuel Oil

2000
    - Coal
    - Refined Hydrocarbons
    - Natural Gas
    - LPG
    - Nuclear Fission
```

FIGURE 6.2 Energy sources at the beginning and the end of the twentieth century. NOTE: LPG = liquefied petroleum gas.

telling" to new heights. Yet, it is my contention that if carbon emissions are to be mitigated or managed, the auto and the energy industries will have to work together to make this prediction a reality. The concept of a renewable hydrogen economy and, specifically of hydrogen being used for transportation energy offers the promise of sustainable auto-mobility.

One method for reducing carbon emissions is by improving the efficiency of future vehicles and propulsion systems. However, this statement begs the question, What will become the dominant engine technology in the next century? Only the marketplace can answer this question, but several clues on possibilities for future engine technologies can be obtained from the efforts of the auto industry-government program Partnership for a New Generation of Vehicles (PNGV).

There are three research goals (Figure 6.4) for the PNGV program, which is looking at the most advanced technologies aimed at improving the energy efficiency of the transportation fleet. The first is the development of new manufacturing techniques to reduce the costs of advanced automotive technologies and improve the efficiency of vehicle manufacturing operations. The second is to develop and introduce new fuel efficiency and emissions reduction technologies on current design vehicles as quickly as possible. The third, and most ambitious-goal is the one most people focus on: to develop a new class of vehicle that provides up to three times the fuel efficiency of today's family sedan. The term "comparable vehicle" means that in addition to providing three times the fuel efficiency of today's cars, this new class of vehicle must also provide the same performance, load-carrying capacity, and range. These conditions must be in place while meeting applicable safety and emissions standards at the adjusted cost of a 1994 midsize sedan.

To be able to demonstrate vehicle prototypes that meet the third goal of the partnership by 2004, the PNGV program embarked on a plan entitled "Invent on Schedule." (Figure 6.5). The first phase of this plan called for compiling the technologies that might contribute to meeting program goals. The second phase focused on eliminating those technologies that were clearly not going to be practical during the "agreed-to" time period of the program. The output of this process has led to the identification of several key technologies that are being pursued with great effort and intensity. The next phase of the program was directed at the development of several vehicle prototypes that have been presented to the public this

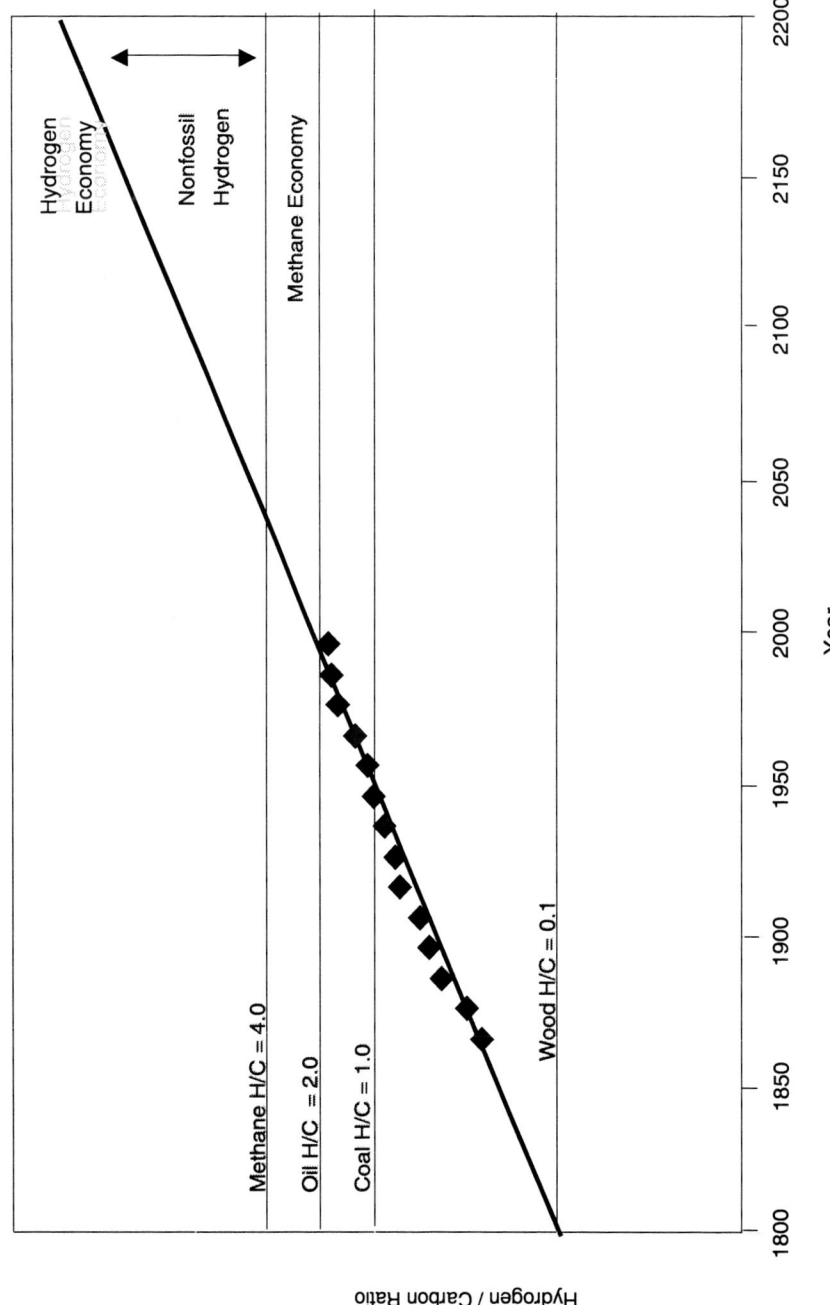

FIGURE 6.3 The trend in hydrogen-to-carbon ratio in global energy consumption. NOTE: y-axis is logarithmic. SOURCE: Nakicenovic, N. (1996) Freeing energy from Carbon. *Daedalus* 125(3): 95-112.

> **Manufacturing**
>
> Reduce manufacturing production costs and product development times for all car and truck production
>
> **Near Term: Conventional Vehicles**
>
> Pursue advances that increase fuel efficiency and reduce emissions of standard vehicles
>
> **Long Term: Next-Generation Vehicle**
>
> Develop a new class of vehicle with up to three times the fuel efficiency of today's comparable vehicle

FIGURE 6.4 Research goals for the PNGV program.

past year. Although these prototypes are impressive examples of automotive technology, they do not meet all of the goals of the program. They indicate that progress has been made as a result of this industry-government partnership and where additional improvements are needed to develop a production prototype in 2004.

During the first two phases of the PNGV program, considerable use of analytical models helped define the expected fuel efficiency of different vehicle-powertrain-fuel combinations (Figure 6.6). The baseline fuel economy calculations in this figure correspond to a 1994 midsize sedan equipped with a port fuel-injected, spark ignition engine operating on a reformulated gasoline. Various improvements in fuel economy can be obtained through use of lightweight bodies, advanced spark ignition engine technologies such as direct injection, and advanced transmissions. The upper and lower ranges for each

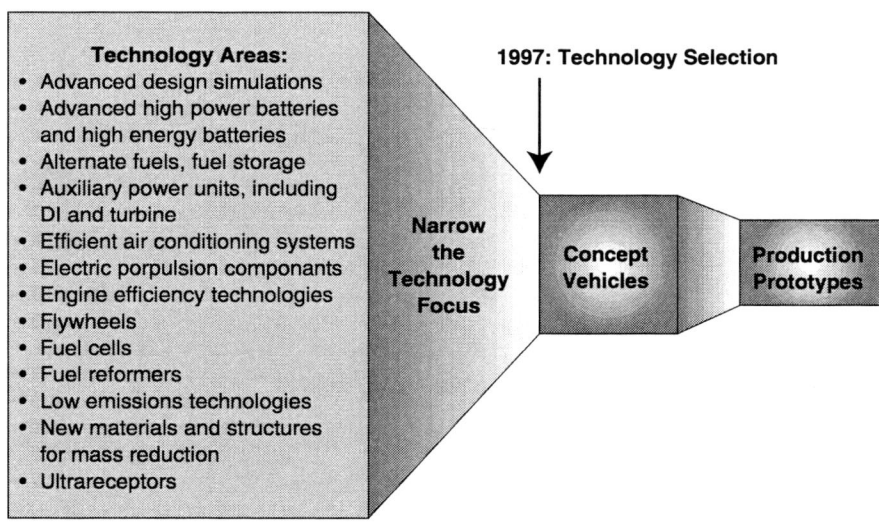

FIGURE 6.5 The "invent on schedule" plan for the PNGV program.

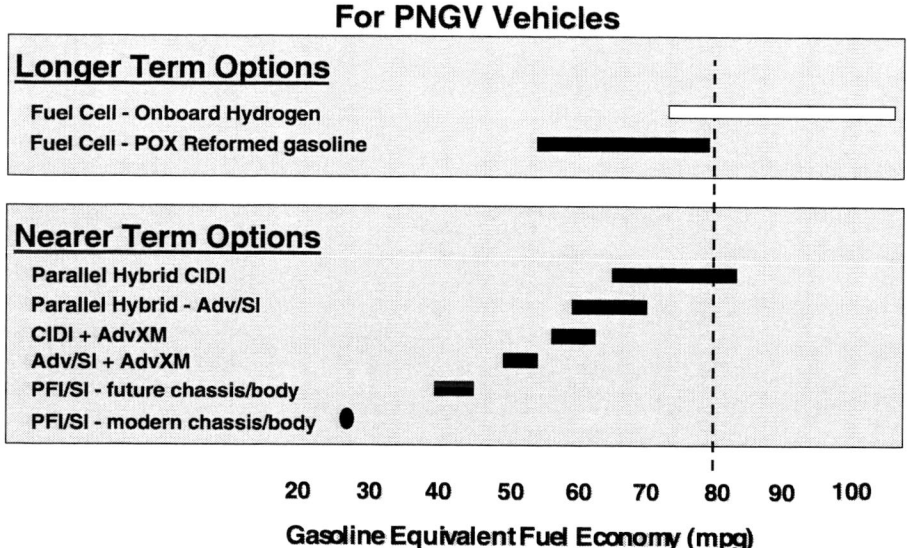

FIGURE 6.6 Fuel economy projections of different PNGV technologies. NOTE: CIDI = compression-ignited direct injection engine, MPG = miles per gallon, PFI = port fuel injected, POX = partial oxidation, SI = spark ignited, XM = transmission.

bar in this figure are based respectively on aggressive or conservative assumptions regarding the efficiency to be gained with each technology. Compression ignition engines provide a substantial increase in fuel efficiency, particularly when installed in hybrid powertrain configurations. In a separate category from internal combustion engines are fuel cell-powered vehicles operating on either gasoline or hydrogen. The results of such modeling efforts have led the PNGV program to identify key promising technologies that must be developed if the program is to meet its stated goals.

The four technologies critical to the program success are shown in Figure 6.7; these include lightweight materials, four-stroke direct injection engines (4SDI), electric traction systems, and proton

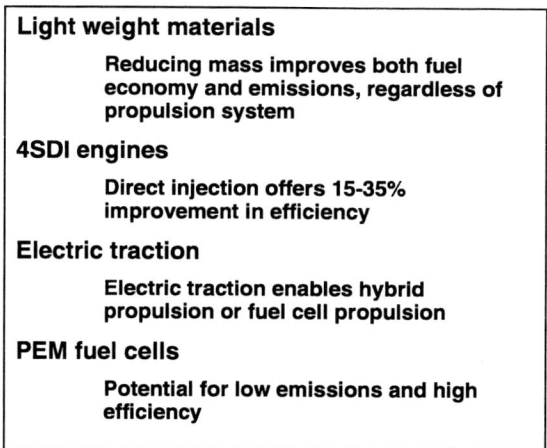

FIGURE 6.7 Four technologies critical to the success of the PNGV program.

exchange membrane (PEM) fuel cells. Although not explicitly stated, it should be clear that the fuel needed for both 4SDI engine and fuel cell operation is a critical aspect of the development of these two propulsion technologies.

A list of the challenges associated with these propulsion technologies is shown in Figure 6.8. Because it has been a viable, fuel-efficient power source for many years, compression ignition engines have been identified as the leading 4SDI configuration. However, challenges remain regarding meeting future emissions standards, particularly for NO_x and particulates, achieving cost targets, and identifying fuel compositions that provide the best fuel efficiency-emissions trade-off. Fuel cells are clearly a riskier propulsion choice. Significant questions remain regarding cost, ability to reform fuels on-board the vehicle to produce hydrogen, which fuel to use, complexity, efficiency, packaging, and mass of the final system.

Given that fuel composition can affect the ability of engine or fuel systems to meet fuel efficiency and emissions goals, it is worth considering what fuel characteristics will be needed to support the introduction of advanced engine technologies (Figure 6.9). Generally, there are several key characteristics that automotive engineers will always want in any fuel. First, it should have the highest energy density possible for maximum range and customer convenience. For internal combustion engines, it should provide good combustion characteristics—that is, the proper octane for spark ignition engines, and high cetane for compression ignition engines. The fuel should have a low tendency to form deposits throughout the fuel, reformer, combustion, and exhaust emissions control systems. The fuel should have no instantaneous or long-term deleterious effect on pre- or aftertreatment catalyst systems. And finally, the fuel should have a reasonable cost.

Within the context of these general characteristics, the nature of a specific fuel depends greatly on the engine technology in which it will be used. For example, in the case of compression ignition engines (Figure 6.10), mechanical and electrical control technologies such as common rail fuel injection systems, direct electronic fuel injection with rate shaping, electronically controlled turbocharging and exhaust gas recirculation, and lean NO_x catalysts and particulate traps will be required to meet emissions goals. These technologies will require fuels that contain virtually no sulfur (certainly less than 15 parts per million [ppm]), reduced density, lower aromatics, potentially an oxygenate component, and increased use of lubricity additives for good long-term durability.

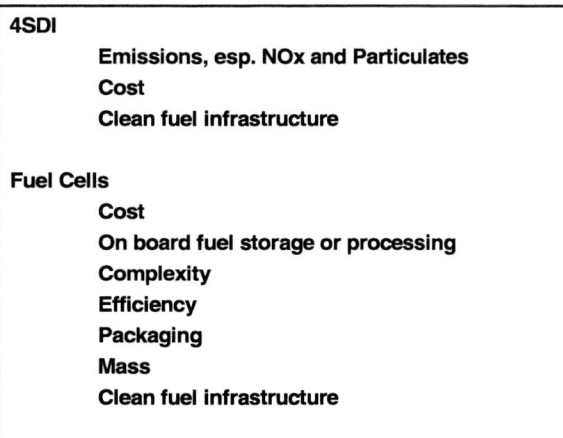

FIGURE 6.8 Challenges associated with four-stroke direct injection and fuel cell propulsion technologies.

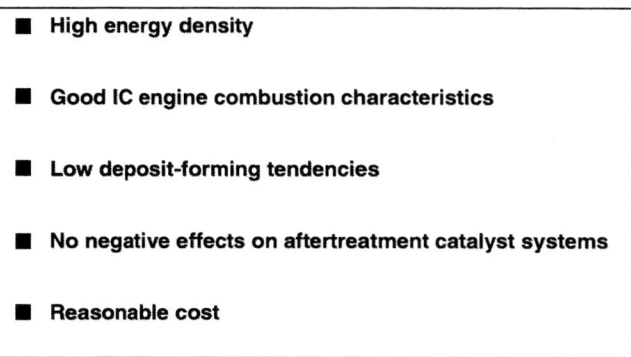

FIGURE 6.9 Desired characteristics for transportation fuels. NOTE: IC = internal combustion.

In contrast as shown in Figure 6.11, for vehicle powertrains that employ a PEM fuel cell and a liquid fuel reformer to produce the required hydrogen, liquid fuels will be required to have zero sulfur (this means less than 10 ppm in the fuel, possibly coupled with an on-board sulfur trap resulting in less than 0.5 ppm delivered to the reformer or stack), as well as the highest H/C ratio possible and an acceptable energy density. There will be no octane requirement, and vapor pressure limits will be needed only to control evaporative emissions from the fuel tank. Some oxygenated components in the fuel blend might be used for a variety of performance attributes, but there cannot be any contaminants that would interfere with or poison the fuel cell stack.

Based on these fuel attributes and engine requirements, the engine-fuel combinations shown in Figure 6.12 are the most likely to achieve substantial market penetrations in different global markets at different times during the twenty-first century. Although each combination is technically possible during the next century, the extent to which any of these combinations will be successful in the marketplace depends on a combination of social, economic, and/or regulatory events.

In the case of spark ignition engines, the best options for fuels include highly reformulated gasoline with virtually no sulfur; alcohols derived from biomass; natural gas; or as the infrastructure develops, hydrogen.

FIGURE 6.10 Advanced engine and fuel requirements for reduced emissions from compression ignition engines. NOTE: EGR = exhaust gas recirculation.

- No sulfur (e.g., < 0.5 ppm)
- High H/C ratio
- Acceptable energy density
- No octane requirement, RVP limits for evaporative emissions control
- Oxygenated fuel component
  - Reduced CO formation
  - Reduced coking
  - Reduced heat loss
  - Quicker starts
- No metal impurities or contaminants (e.g., < 0.5 ppm chloride)

FIGURE 6.11 Fuel attributes for good on-board fuel processor operation. NOTE: RVP = vapor pressure.

For compression ignition engines the most promising fuels include highly reformulated diesel fuel with virtually no sulfur; liquid hydrocarbons derived from natural gas; diesel fuels blended with specific oxygenate components; alcohols derived from biomass; or natural gas possibly blended with some percentage of hydrogen.

- Spark Ignition, Lean Burn Engines
  - Highly reformulated gasoline
  - Alcohols from biomass
  - Natural gas
  - Hydrogen

- Compression Ignition Engines
  - Highly hydrocracked distillates (reformulated diesel)
  - Fischer-Tropsch hydrocarbons from natural gas
  - Oxygenate blending components
  - Alcohols from biomass
  - Natural gas/hydrogen blends

- Fuel Cell Electrics
  - Light naphtha
  - Fischer-Tropsch hydrocarbons from natural gas
  - Hydrogen

FIGURE 6.12 Potential engine-fuel combinations for the twenty-first century.

Finally, the best alternatives in the near term for fuel cell electric vehicles include light naphtha or alkylate refinery streams; liquid hydrocarbons derived from natural gas; or as production, storage, and distribution systems develop, hydrogen.

The desire to "manage" carbon emissions as future vehicles are added to the global car park suggests that some of these listed options are more desirable than others. If, in addition to meeting engine or propulsion system requirements and customers' expectations, we add the goal of no net CO_2 emissions from the vehicle fleet, then the best long-term options in my perspective are renewable hydrogen or biomass-derived ethanol.

I have not included the processing of fossil fuels to produce hydrogen coupled with "decarbonization" or sequestration facilities at centralized plants for capturing solid carbon or CO_2 emissions. Technology does exist for capture of these H_2- manufacturing plant by-products, and several strategies have been suggested for their disposal. Although such technologies could be a possible near-term mechanism for lowering CO_2 emissions and speeding introduction of H_2, both decarbonization and sequestration strategies suffer from significant drawbacks. Collection and disposal of such by-products add substantial effort, inefficiency, and cost to the production of transportation fuels that are avoidable if renewable, non- or low-carbon fuels are produced in the first place.

It's worth reviewing the importance of each of these options. In the case of hydrogen, I've already pointed out that it could turn out to be the "ultimate" fuel. Today's fuels are increasing in hydrogen content, hydrogen can be made without generating any CO_2 emissions, and I believe the energy industry would be receptive to the use of hydrogen since it would represent the last fuel change needed.

In addition, hydrogen enables a variety of advanced engine or propulsion technologies including "zero-emission" fuel cell vehicles, simplified fuel cell vehicle designs without the need for a fuel reformer, and internal combustion engines that have both improved efficiency and very low emissions (Figure 6.13).

The concept of hydrogen-powered fuel cell vehicles is far more than a futuristic prediction. General Motors introduced a significant prototype minivan at the 2000 Geneva Motor Show that operates on cryogenic hydrogen fuel. More than an auto show concept, the Zafira minivan (Figure 6.14) is a fully operational, five passenger vehicle that has been demonstrated in five different countries, including pacing the men's and women's marathons at the Sydney Olympics.

Although the current Zafira is designed to utilize liquid hydrogen, General Motors is undecided on an optimum form of hydrogen storage for future vehicle applications (Figure 6.15). We are currently investigating storage methods that utilize compressed gas, cryogenic liquid, and solid adsorption tech-

- ■ Hydrogen (H_2) recognized as the ultimate fuel
  - Trend toward increased H_2 content fuels
  - Could be made without CO_2 emissions
  - Energy industry receptive because H_2 would be last fuel change needed
- ■ H_2 enables advanced vehicle technology
  - H_2 powered fuel cell vehicles qualify as ZEV's
  - H_2 storage allows simplified vehicle design (no reformer)
  - H_2 IC engines have improved efficiency, near-zero emissions

FIGURE 6.13 Why hydrogen is important to General Motors. NOTE: ZEV = zero emission vehicle.

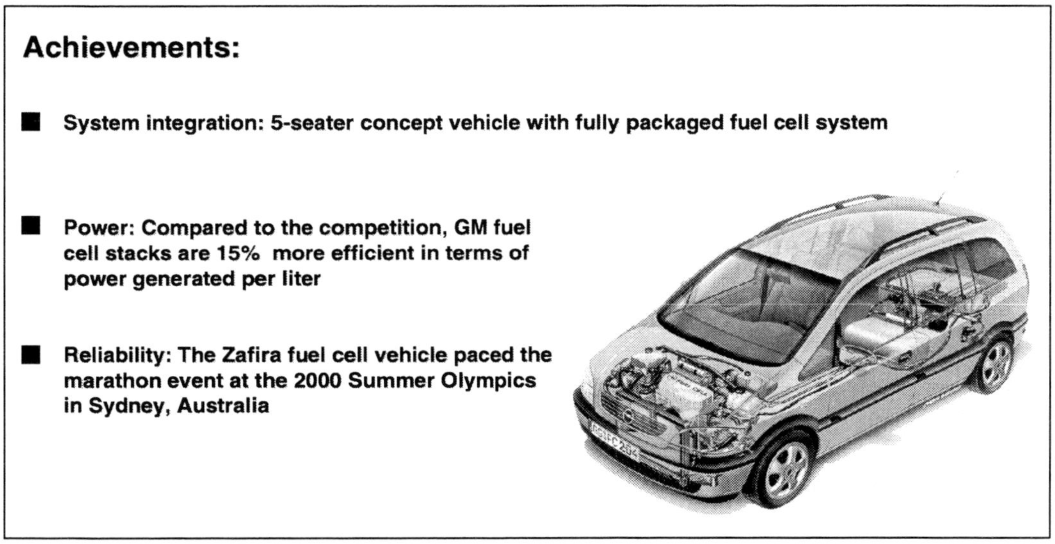

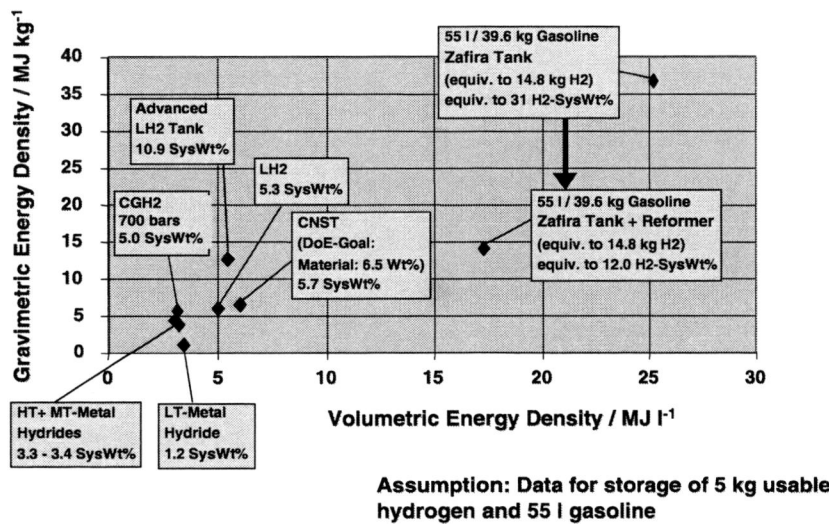

FIGURE 6.14 The hydrogen-powered Opel Zafira at the Geneva Motor Show 2000.

nologies. We are also comparing the efficiencies and utility of such storage methods with gasoline reformer-equipped vehicle designs. It's claimed that advanced tank designs for both gaseous and liquid storage, as well as advanced solid adsorption materials, can store between 5 and 10% hydrogen calculated as a fraction of the overall weight of the total fuel storage system. Such technologies still store substantially less energy than is possible from liquid hydrocarbon fuels, even when the weight of the reformer system is added in. Advanced materials research is needed to improve both the materials of construction for gaseous and cryogenic tanks and new solid formulations for adsorbing greater amounts of hydrogen.

FIGURE 6.15 Energy storage comparison: gravimetric versus volumetric energy density of hydrogen storage.

To demonstrate that solid adsorption hydrogen storage systems are technically feasible, a version of the Precept (Figure 6.16), GM's PNGV technology demonstration vehicle, was equipped with a fuel cell and a low-temperature hydride storage system. It is useful to note that the gasoline-equivalent fuel economy for this vehicle is calculated as 108 miles per gallon (mpg). Although successful as a developmental prototype, substantial improvements in many aspects of fuel cell vehicle design are needed prior to any future production vehicle program.

Developing a suitable hydrogen storage system is only part of what it takes to create a hydrogen transportation fuel industry. The other necessary components are the production technology used to create renewable hydrogen (Figure 6.17) and the infrastructure to deliver it to the driving public. The generally accepted technique for producing hydrogen today is through steam reforming of natural gas. In the short-term, this process can be coupled with decarbonization techniques to produce hydrogen with little CO_2 generation. In the longer term, however, this technology could be transferable to the production of hydrogen using biomass-derived or sewage- and garbage-derived methane. Carbon dioxide generation could be substantially reduced if not brought to a CO_2-neutral balance point through the use of biomass-derived hydrogen.

Longer-range hydrogen generation techniques could include both electrolysis of water using renewable electricity and direct generation of hydrogen through use of bacteria and algae systems. Both of these technologies require significant development to further improve generating efficiencies and reduce production costs.

The other critical component of a hydrogen fuel transportation system is an infrastructure designed to compete with other forms of energy (Figure 6.18). This can be achieved if we provide value to the driving customer by creating a distribution and storage system that is safer than gasoline, has widespread availability, meets customer expectations for value, and is easy to use. There is no room for compromise on any of these issues. Gasoline and diesel fuels are very good energy sources for transpor-

FIGURE 6.16 The Precept at the Detroit Motor Show 2000.

- Renewable methane
- Water electrolysis with renewable electricity
- Direct generation of biomass-derived hydrogen

FIGURE 6.17 Necessary ingredients to create renewable hydrogen.

tation systems. They meet many of the criteria that automotive engineers desire in a fuel. Customers will not accept a new fuel unless it provides the same or greater levels of safety, convenience, and value.

In the end, the use of hydrogen in transportation applications will depend as much on research and development of production and delivery technologies as any issue associated with hydrogen use on-board the vehicle.

The other fuel option identified as a possibility for management of carbon emissions is biomass-derived ethanol. The concept of biomass-derived ethanol can be focused further to ethanol derived from cellulose (Figure 6.19). Currently the ethanol used for blending with gasoline in many states in the United States is derived primarily from glucose (starch) in the corn kernel. General Motors does not believe that a large percentage of U.S. transportation fuel needs could be produced from this source of biomass. Furthermore on a societal basis, we do not believe that transportation energy needs should compete with human food production. Thus, we conclude that ethanol used for transportation needs should be derived from cellulose.

Cellulose-derived ethanol could provide a source of fuel that would consume as much CO_2 as it produces during the combustion process, thus making the transportation sector CO_2 neutral. Technologies for the production of ethanol for cellulose are known, but they are not cost competitive with production from glucose. Infrastructure changes are needed to minimize contamination of ethanol by water-soluble components, but the changes would be of less magnitude than for introduction of a hydrogen infrastructure. Ethanol can be produced from the cellulose of renewable crops such as switch grass and poplar trees. Calculations have shown that there is sufficient land available in the United States to replace the 100 billion gallons of gasoline used each year, thereby also providing substantial improvements in U.S. energy independence.

From a vehicle perspective, the introduction of fuels containing large amounts of ethanol is a much simpler task than producing hydrogen-fueled vehicles. As allowed by current law, several manufacturers have introduced E-85 vehicles into the marketplace in exchange for limited corporate average fuel

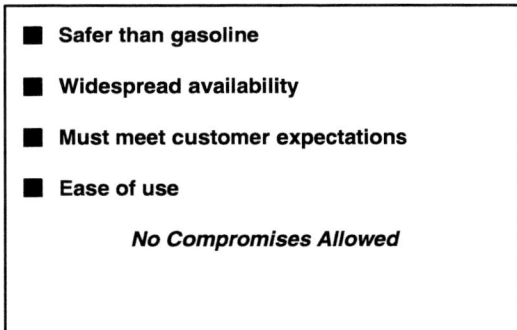

- Safer than gasoline
- Widespread availability
- Must meet customer expectations
- Ease of use

No Compromises Allowed

FIGURE 6.18 Necessary ingredients to create a hydrogen infrastructure.

- **Potential for CO$_2$-Neutral, Transportation Fuels**
  - Technology known, but not cost competitive
  - Infrastructure changes needed, but not the same magnitude as those for hydrogen
  - Renewable; energy independence
- **Ethanol-Compatible Vehicle in Production**
  - GM will have over a million E85 vehicles on the road by 2004
  - Vehicle modifications relatively less expensive than for other alternative fuels

FIGURE 6.19 Ethanol derived from cellulose: its importance to General Motors.

economy (CAFE) credits. GM will have more than a million E-85 vehicles on the road by 2004. These vehicles are designed to operate on 100% gasoline or any blend of ethanol and gasoline up to 85% ethanol. Admittedly, most of these vehicles currently operate on straight gasoline, but that's only because the number of E-85 stations is limited. The modifications of an E-85 vehicle are relatively less expensive than those for alternative fuel candidates.

An example of GM's current E-85 vehicle offerings is the Chevrolet S-10 pickup truck. E-85 versions of the Chevrolet Suburban and Tahoe and GMC Yukon and Yukon XL sport utility vehicles equipped with the 5.3 liter V-8 have been added in model year 2001.

Finally, it is worth noting that calculations conducted at Argonne National Laboratories have demonstrated that it is possible to reduce petroleum consumption by approximately 90% and essentially eliminate all of the CO$_2$ generated by mobile sources through use of ethanol generated from cellulose (Figure 6.20). The range of estimates depends on whether herbaceous or woody biomasses are being used as a source of cellulose. Reductions in greenhouse gases can exceed 100% due to natural, long-term carbon sequestration in the soil. Given the installed capital base associated with refineries and petroleum pipelines, converting the transportation system to an alternative fuel such as ethanol derived from cellulose would require a number of years. If managing carbon is decided to be in the national

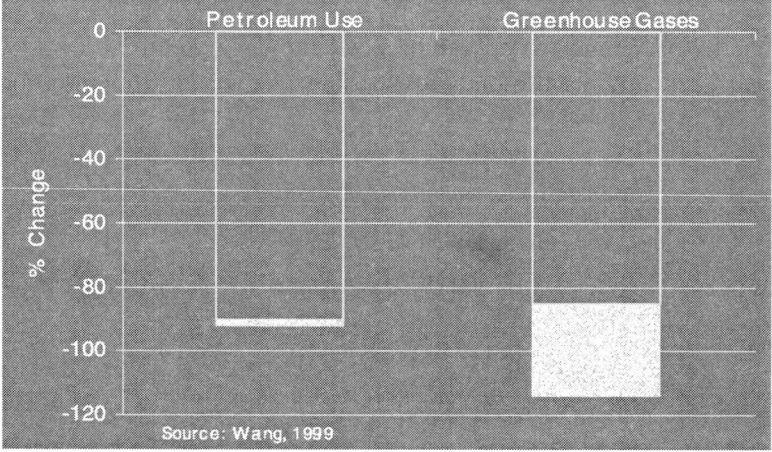

FIGURE 6.20 Potential future impact of ethanol from cellulose. Adapted from M. Wang, C. Saricks, and D. Santini (1999). Effects of fuel ethanol use on fuel-cycle energy and greenhouse gas emissions, Argonne National Laboratory Report ANL/ESD-38, January 1999.

interest, ethanol from cellulose needs to be given serious consideration as a technologically achievable solution.

In summary, General Motors believes that carbon management will be a major driver for development of future engine-fuel systems. It is expected that future transportation fuels will continue the trend toward lower density and higher H/C ratios. Reformulated gasoline and diesel fuels will provide the opportunity to reduce CO_2 generation through improved internal combustion engine efficiency meeting applicable emissions standards. The best longer-range options for achieving a CO_2-neutral state include (1) renewable hydrogen and (2) ethanol derived from cellulose. Both additional research efforts and a national long-range energy plan (longer than one administration) are needed to achieve future carbon management objectives.

DISCUSSION

Richard Wool, University of Delaware: Significant advances have been made in genetically engineered plant oils, and at the same time, some advances have been made with bio-diesel. Have you looked at this possibility?

James Spearot: Bio-diesel has some interesting aspects associated with it, but right now it is not cost competitive. If we decide as a nation and as a global society that we need to control carbon—that wasn't the point of my talk here—then we have to go far beyond bio-diesel.

Now, let me say something about genetically modified crops. I think this is a technology that deserves its day in the sun. Unfortunately, I am concerned that it will run into some of the environmental concerns that have been generated with other advanced technologies. I hope this doesn't happen. We would argue that grasses could be grown for fuel production and kept out of the food system, but unfortunately, as you know, once you put things out into nature, it is difficult to keep them segregated.

David Keith, Carnegie Mellon University: It is clear that there is great potential for biomass. Current estimates put the price of biomass at about $2.50 per gigajoule of primary energy. By comparison, the price of energy, particularly if we require liquid fuels, from other non-fossil renewables is substantially higher. Yet I also want us to think hard about the environmental impact of biomass use, because this is the metric that counts, not just CO_2 or renewability.

Consider the fact that, in the United States, where we eat a lot of meat, it takes roughly a hectare per person to feed ourselves and we use 10 kw of primary power each. The largest energy flux you get out of biomass is a watt or two per square meter, which equals about a hectare per person. So, that means if you wanted to use biomass for the third of our energy consumption that is used in transportation, you need increase our current total cropland use by about by about one-third. I think we must think hard about whether we want this future.

Arguably, our land use is the biggest single environmental impact we make. You need only fly over the middle part of this country to see the level of current impact. Think about what a substantial increase in land use would mean. So, I realize that this is not really a fair question for you because it is not GM's responsibility to grapple with these problems, but I would like you to comment on the potential of biomass.

James Spearot: You helped me answer the question with your last sentence. Your question was, What are our options for transportation fuels? I believe that ethanol derived from cellulose is a viable option that should be looked at. Clearly, there is no option that is obvious or easy. Every option has pluses and minuses associated with it, and you have identified some of the minuses of deriving liquid fuels from agricultural crops.

I think it has been shown that there is enough land available, but it is fair to ask the question of whether we want to devote all that land to switch grass or poplar trees. This is something that society as a whole will have to make a decision on when the time comes.

Alex Bell, University of California at Berkeley: The hydrogen fuel cell has always been an attractive option for automotive driving force. The issue is one of materials of construction. When I was serving on the Partnership for a New Generation of Vehicles Review Committee, it became apparent that with the current loadings of precious metals required for the fuel cell, even excluding the reformulation apparatus, you could probably build on the order of a million cars a year. This is one-twentieth of what we produce in the United States alone.

So, what is the long-term viability of looking at hydrogen fuel cells for the world, given that the resources of these precious metals don't reside here in the United States?

James Spearot: We are now working on our sixth-or seventh-generation fuel cell stack. We have lowered the noble metal content per power output on every generation. In fact, the numbers are coming down dramatically.

I can't quote the numbers from memory, but I wish the cost, the size, and the complexity of the fuel reformer were coming down as quickly as the efficiency of the fuel cell stack is improving. I think we will get to the point where we can have a manageable amount of noble metal in the fuel cell. It may be that the first few fuel cell vehicles sold are over-designed in terms of noble metal usage, but we believe that fuel cells are a viable option. We are going to have to solve that problem if we want to produce these vehicles.

That is the approach we are taking to this particular issue, but there is a critical need for additional electrochemical research to develop fuel cell systems that have reduced amounts of noble metal. We are lowering the amount, but it is not where it needs to be yet.

Tobin Marks, Northwestern University: Let me make sure I understand. The ethanol from cellulose, what is the process for that? Is it fermentation or some other process?

James Spearot: Can any of the life science people answer this? I am a chemical engineer with a background in gasoline and refinery operations. My understanding is that it consists of a variety of processes put together. They are well-known processes. It is just that they are more expensive and energy intensive than making ethanol from corn. My limited understanding is that there is significant research to be done on enzyme processes to speed the organic chemistry going on.

Tobin Marks: Many are not optimized yet.

James Spearot: Well, they have been working on optimizing it for a long time, but the hope is that they can still make further improvements.

Klaus Lackner, Los Alamos National Laboratory: The issue of whether to produce hydrogen from electrolysis or through some fossil fuels, where you then have to go through the sequestration side, ultimately boils down to a cost issue. For comparison, if you take gasoline at $1.50, there are several hundred dollars per ton of CO_2 right in the cost of the fuel. So if you now estimate $30 per ton of CO_2

for sequestration, this is on the order of a 10% correction. If you start with electricity as the basis for making hydrogen, you start with a very expensive source of energy. Since there is a finite efficiency in that process, the total amount of energy consumption is even higher. No matter which way you go, electrolysis is an expensive way of making hydrogen.

James Spearot: In some parts of the world, there is cheap hydrogen available where there are hydroelectric capabilities. I am one of the people who believes that eventually we are going to have to look at additional nuclear energy sources in the future. In fact, I am concerned that we might be falling behind countries such as France and Korea, which are very interested in nuclear energy.

So, today, you are absolutely right. I don't argue with where we are today, but again we are looking out over a period of time. I don't even mind starting from producing hydrogen from steam reforming with sequestration as a bridge technology to get the hydrogen infrastructure in place. I think that is a viable option, but I think, in the long term, we have to figure out how to make renewable hydrogen in a cost-effective manner. That is the only point I am trying to make.

Brian Flannery, Exxon Mobil: We have certainly looked at fuels from biomass, which are often promoted as being potentially net CO_2 free, but in current practice, they are very far from it. In fact, they are virtually identical to using fossil fuels directly because of the enormous energy that goes into the cultivation, harvesting, refining, and processing, et cetera.

So, I think an option that needs to be explored very carefully is real systems and how they could work in practice. We need to find out whether a system has the potential to be as effective as some ultimate optimum potential might be.

However, the use of biomass does raise huge questions of land use and other environmental impacts that I think need to be on the table along with the climate change issues and the environmental impacts that may be associated with that.

James Spearot: I don't disagree with anything you have said, Brian. I will argue only that the economics and the CO_2 savings that we are looking at today are based on ethanol from corn and ethanol blended primarily to 10% in some gasoline. I think the economics will change dramatically with different crops and with larger amounts of ethanol use, as will the amount of CO_2 that is saved.

However, it is one of the options and I believe it ought to be on the table. I guess we can agree to disagree because I believe sequestration has some significant environmental issues associated with it also.

Alan Wolsky, Argonne National Laboratory: I recollect estimating that using today's technology and today's electricity prices, hydrogen from electrolysis of water was $11 per million British thermal units. That is expensive. I am not aware of any nuclear advocate who thinks that electricity from nukes is cheaper than electricity from gas turbines.

James Spearot: I don't have figures to refute anything you have said. I would argue only that as we go forward during the next century, there is a reasonable hope that technology will bring hydrogen production cost down.

The German government believes that hydrogen is, in fact, competitive today with gasoline in Germany at $4 a gallon gasoline price. I don't know whether this is right or not. I don't know where they are planning to get their hydrogen from, but they have a very real interest in the use of hydrogen in Germany.

In fact, I suspect that if GM produces fuel cells at some point in time, we will see them in Europe before we see them in the United States because of the higher fuel costs there.

John Stringer, Electric Power Research Institute: The cost of electricity from nuclear plants, as I said before, depends very much on the price of capital. If the capital cost is low enough, then the electro-steel research science is less, but that is not the point. The point is that if you have a nuclear station and you are able to run at base load all the time, then the comments that you made are quite right.

However, if—and this is usually the case—the plant doesn't go on at base load all the time, the period when it isn't going at base load can be used to generate hydrogen at a low marginal cost. That is the issue, I think.

Now, how that works out for an economy requiring hydrogen in large quantities is a matter of the scenario in which you put it, but there are scenarios in which you can get very cheap hydrogen by use of marginal hours from nuclear power.

Tom Rauchfuss, University of Illinois: I have a question about what progress you have made on high-density storage of hydrogen. Are there new materials that perhaps store more hydrogen than palladium, which is maybe 1%.

James Spearot: Unlike many of the reports that are surfacing all over the world, we have not been very successful in identifying materials that absorb large amounts of hydrogen. We are still working on this particular issue. I don't want to discredit anybody's work, but we are concerned about some of the reports because we are not able to reproduce some of the work that is being done, and we are trying very hard to reproduce it.

The most positive news we have in terms of storage is some of the numbers that are currently being evaluated in prototyped, compressed-hydrogen tanks. Right now, tanks with capability of 5,000 pounds per square inch (psi) are almost validated, and tanks with a capacity of 10,000 psi are being looked at, although they are a long way from being validated. With these two technologies, you can begin to achieve 5 or 6% hydrogen storage range.

Participant: When we talk about producing energy from agricultural sources, whether it be biomass or anything else, we have to take into consideration the energy balance between growing, harvesting, and transporting these materials so that they can be converted, let's say, to ethanol. This energy balance has been negative in the past. Is there a change now?

James Spearot: I can't answer your question specifically with numbers. All I can tell you right now is that our analyses have been based on total life-cycle energy and materials costs, and based on these total life-cycle analyses, we believe that ethanol from biomass is competitive. Ethanol from cellulose is more expensive than petroleum-based energy forms today, but there is a very real reason to think that improvements can be made in cellulose processing techniques that would allow us to bring the cost of the ethanol derived from cellulose down to a competitive level.

In terms of CO_2, the life-cycle analysis has said that there are benefits from utilizing ethanol derived from cellulose, and work done at Argonne supports those numbers.

7

Renewable Energy: Generation, Storage, and Utilization

John Turner
National Renewable Energy Laboratory

National security depends on energy security.
—President-elect George W. Bush

Energy is the major input for overall socio-economic development.
—C. R. Kamalanathan, Secretary, Ministry of Non-Conventional Energy Sources, Government of India.

The Americans in this area are very much the villains of the piece. They've not gone along with Kyoto and yet they are unquestionably the largest polluter with 4% of the world's population and 25% of greenhouse gas emissions.
—Sir Crispin Tickell, the former British ambassador to the United Nations.

For our own security we must reduce our dependence on foreign sources of energy. However, for our economy to grow we must obtain additional energy. Developing countries will also compete for this energy, because they must have additional energy sources before they can raise their standard of living and join the world's developed nations. Like a dark cloud hanging over all of this are the pollution and global climate change these fuels produce. As we use more coal, oil, and gas, we only exacerbate the problem.

As we seek to address the issues of global climate change and carbon management, we must remain aware of all these other issues. To control and sequester carbon emissions will require additional energy. This will put additional pressure on our fossil reserves, our supply lines, and the energy infrastructure.

Currently, this world is inhabited by more than 5 billion individuals and is powered ultimately by solar energy. The food we eat and the oxygen we breathe comes from photosynthesis. Can we use this energy from the sun that indirectly powers our bodies to provide the energy we need to run our society? This chapter discusses the possibilities and offers some suggestions as to the pathways to a sustainable energy infrastructure.

Sustainable in this context means capable of supplying a growing population with energy without destroying the environment within which it is used. It must also include the ethics of using the earth's resources, particularly fossil fuels. It is completely unethical for us to squander the finite nonreplaceable resources of the planet in a one-time use without any accountability to future generations and our environment. We must try to look ahead, envision a future society, give it a voice, and use this to point out those issues that would clearly affect future lives. To consume a resource without developing a replacement is clearly an issue that will affect future generations.

The author's perspective in writing this article include the following:

- To modify an existing energy infrastructure or build a new energy infrastructure requires money and energy—energy that must come from existing resources.
- Advanced renewable energy systems can provide long-term benefits to society—namely, sustainability.
- Manufacturing renewable energy systems for the developing world provides an economic benefit to the United States because a very large portion of the energy demand will occur in these regions. The points to consider here are the following:
  1. What is the market for renewable technologies versus sequestration technologies?
  2. Distributed generation that uses indigenous local resources reduces the need to build and maintain large electrical grids.

SEQUESTRATION VERSUS RENEWABLES

Our current energy and transportation infrastructure is plagued by many problems. It pollutes and damages our environment, it makes us dependent on foreign governments for more than 50% of our supply, and the trade deficit resulting from our oil purchases (at more than $1 billion per week) has a destabilizing effect on our economy. In addition, we must maintain an enhanced military presence in the Middle East to keep our access to this oil, which puts our military at risk. If we implement large-scale sequestration, it solves only the environmental issue, but the rest of the problems are actually exacerbated—we will have to burn more fossil fuel in order to generate the additional energy needed to power the sequestration systems to remove the CO_2 that our fossil fuel systems generate. This means importing more oil (increasing our trade deficit) or burning our own reserves at a faster rate.

Sequestration is only a temporary fix; eventually we will have to replace fossil fuels. Since we have technologies that will "do the job," we should implement a sustainable energy infrastructure that doesn't emit carbon dioxide and can supply all our energy needs using our indigenous resources. Renewable technologies also have the ability to expand as our usage of and needs for energy grow. Furthermore, renewable energy systems can be configured to supply not only all the electrical needs, but also all our transportation requirements.

The United States can spend its money and energy resources building sequestration systems or implementing renewable energy technologies; it is not likely that we can or will do both. The first part of this chapter therefore discusses the vision and possibilities of renewable energy such that the reader will be convinced of the viability of this approach. In the second part, a general list of research areas is presented. Because it would be impossible to cover all of the associated technologies fully, the author has chosen to present those technologies he feels could have an impact in the immediate to 10-year time frame.

FEASIBILITY OF RENEWABLES

We first address the question, Can we really supply all our energy needs from renewable energy? The power of renewable energy can easily be shown using the United States as an example.[1] The United

States is the world's largest energy consumer. Total U.S. annual electrical demand for 1997 was about 3.2×10^{12} kWh (representing 25% of the world's consumption). If we assume flat fixed-plate collectors with a system efficiency of 10% (current commercial technology) covering only half of the employed land area, a photovoltaic (PV) array 104 miles (166 km) on one side (~10,900 square miles or 27,600 km^2), placed in southwest Nevada, would, over one year supply all this energy. This area represents less than 0.4% of the available land area of the United States. A system efficiency of 15%, which should be available in the next three to five years, would decrease the area to 7,200 square miles. If we add wind to the energy mix, this area for PV decreases further. If we add geothermal and hydropower, the area gets smaller still. The point is clear: we can gather more than enough renewable energy to power our society and yet have an abundance of renewable resources available for future growth. Also, one should note that wind alone or solar thermal alone could provide all our electrical energy needs.

Although we have used Nevada for this calculation, PV panels can be placed across the entire United States. The U.S. average solar irradiance is 1,800 kWh/m^2 per year. Implementation would involve the rooftops of homes and businesses, parking lots, and even landfills, along with 1 to 10 square mile "energy farms." Landfills are especially viable. There are more than 10,000 landfills in the United States, and only 3,000 are active. The land is typically unusable for 20 to 30 years, and this is exactly the lifetime of PV systems. Furthermore, landfills produce methane as the organic matter decomposes. If methane is collected and used in a fuel cell, for example, this provides an additional energy input from the land.

The way in which electricity is generated is as important as the area required to supply this amount of energy. If we look at solar irradiance data (how much sunshine is available per day), we see that, in southwest Nevada, the sun shines only for an annual average of about 6 hours a day. That effectively means that this system generates the same amount of electricity in 6 hours that the United States uses in 24 hours. The remaining 18 hours of electricity must be taken from the energy stored during the hours of sunlight. Energy storage and its efficiency then become critical; any efficiency losses must be made up by an increase in the area of the PV array. This points out one of the major drawbacks to many forms of renewable energy—their intermittency and the need for energy storage. Dealing with the intermittent nature of certain forms of renewable energy and energy storage systems are topics in this discussion.

ENERGY PAYBACK

Implementing an energy infrastructure that uses more energy in its manufacture and deployment than it produces in its lifetime is not a viable pathway for the future. In fact, our current energy infrastructure has an energy payback ratio of about 0.3 meaning it converts only 30% of the input energy (in the form of fossil fuel) into electricity. A sustainable energy system will have an energy payback ratio greater than one. There is, however, the persistent myth that it takes more energy to manufacture renewables than they produce in their lifetime. Actual calculations show a very rapid payback. For example, the energy payback time for current PV systems has been calculated to range from three to four years depending on the type of PV panel (thin-film technology or multicrystalline silicon respectively).[2,3] This includes the energy costs for processing the semiconductor material and assembling it into a module, the frame, and the support structure. This means that PV with a lifetime of at least 30 years has a payback ratio of 8 to 10. Wind energy has an even faster payback—two to three months, and this includes scraping the turbine at the end of its life.[4] For wind, with a 20-year lifetime, the payback ratio is an impressive 80! Renewable energy resources can therefore be used to manufacture additional renewable energy systems, like a breeder plant, producing more energy than they use in the manufacture—the ultimate in sustainable energy systems.

With the combination of PV, solar-thermal, wind, geothermal, biomass, and hydro, the United States has sufficient renewable resources for virtually unlimited energy growth. Moreover, PV, wind, geothermal, and biomass are commercial systems that are available now. We now discuss some of the renewable generation technologies and then highlight some of the research issues.

PHOTOVOLTAICS (SOLAR CELLS)

Photovoltaics (solar cells) convert light energy directly into direct current d.c. electricity with no moving parts.[5] Developed in the early 1950s, primarily for the space industry, PV is now a $1.5 billion industry for terrestrial applications. Worldwide production in 2000 was approximately 278 megawatts (MW), representing a yearly growth rate of about 37%. As of January 2001, the production of major manufacturers was sold out for next two years, and many new production facilities are being built. Levelized electricity costs are reported to range from 15 to 30/kWh ($5-$10 per watt installed). The major application in the past has been for remote applications far from the grid, but it is becoming more common to find PV on the rooftops of homes and integrated into buildings (Figure 7.1)

In fact, building integrated PV has become a major market. In these days of heightened awareness of energy shortages, and in particular their effect on Internet businesses, energy reliability is becoming a major factor in the decision of companies to integrate PV into their buildings. The siting issues are very important. PV does not impact the area with emissions as diesel generator sets do, and because there are no moving parts, maintenance is minimal. During the summer months, PV provides its maximum power just when air conditioning loads are the greatest. Integrating PV into a building design decreases the installation costs, provides for additional energy reliability and reduces the load on the local grid since the electricity is generated at point of use.

While the major technology being installed today is based on single and polycrystalline silicon, thin film solar cell technologies offer the potential for very low cost and high volume manufacturing resulting in a levelized cost in the 6 cents/kWh range. Thin film technologies utilize 1-5-μm-thick films of semiconductors on glass or stainless steel.

The advantages that thin films provide are high efficiency, reduced materials requirements, and an inexpensive and rapid manufacturing process. The disadvantages are that current systems use toxic or rare materials and that the long-term stability (>20 years) of current technology is unknown, although data to date show excellent stability. Before these materials can make a major impact in the PV market, improved understanding of the scientific and technological base for today's thin films will be necessary (Figure 7.2).

Figure 7.3 shows the historical trends in laboratory efficiencies for the various thin film systems. For solar cells, the efficiency of commercially produced panels usually lags the laboratory efficiencies by about 10 years. For commercial applications, panel efficiencies need to be in the range of 10%, although many of these thin-film systems can be successfully marketed with efficiencies of 5-8%.

ADVANCED DEVELOPMENT OF SOLAR CELLS

Increasing the efficiency of any solar converter will decrease the area that must be covered to collect a fixed amount of energy. Depending on the cost of the solar converter system, this can also lead to lower costs. For photovoltaics, one can greatly increase the conversion efficiency by designing cells to utilize specific areas (colors) of the solar spectrum and stacking them on top of one another in a series configuration. A commercial example is found in the $GaAs-GaInP_2$ tandem cell currently used to power communication satellites. This solar cell consists of a gallium arsenide (GaAs) bottom cell connected to

FIGURE 7.1 4 Times Square, a 48-story skyscraper at the corner of Broadway and 42nd Street, was the first major office building constructed in New York city in the 1990s. The building's most advanced feature is the photovoltaic skin, a system that uses thin-film PV panels to replace traditional glass cladding material. The PV curtain wall extends from the thirty-fifth to the forty-ninth floors on the building's south and east walls. The developer, the Durst Organization, has implemented a wide variety of healthy building and energy-efficient strategies. Kiss and Cathcart architects designed the building's PV system in collaboration with Fox and Fowle, the base building architects. Energy Photovoltaics of Princeton, New Jersey, developed the custom PV modules.

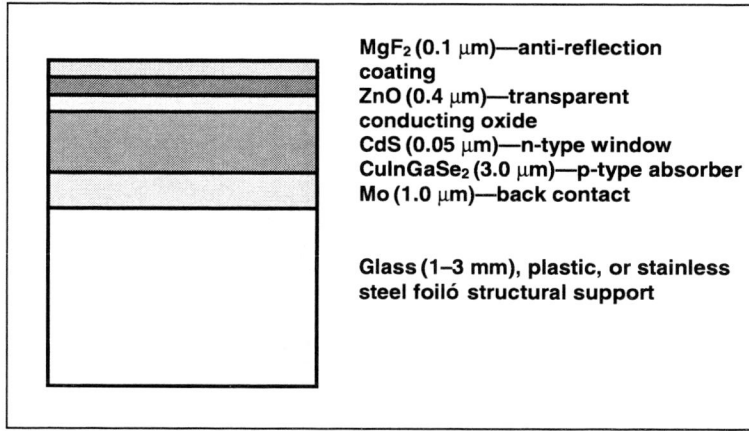

FIGURE 7.2 Thin copper indium diselenide solar cell. Progress in manufacturing is mostly empirical, with little understanding of material properties, devices, and processes that lead to higher efficiency.

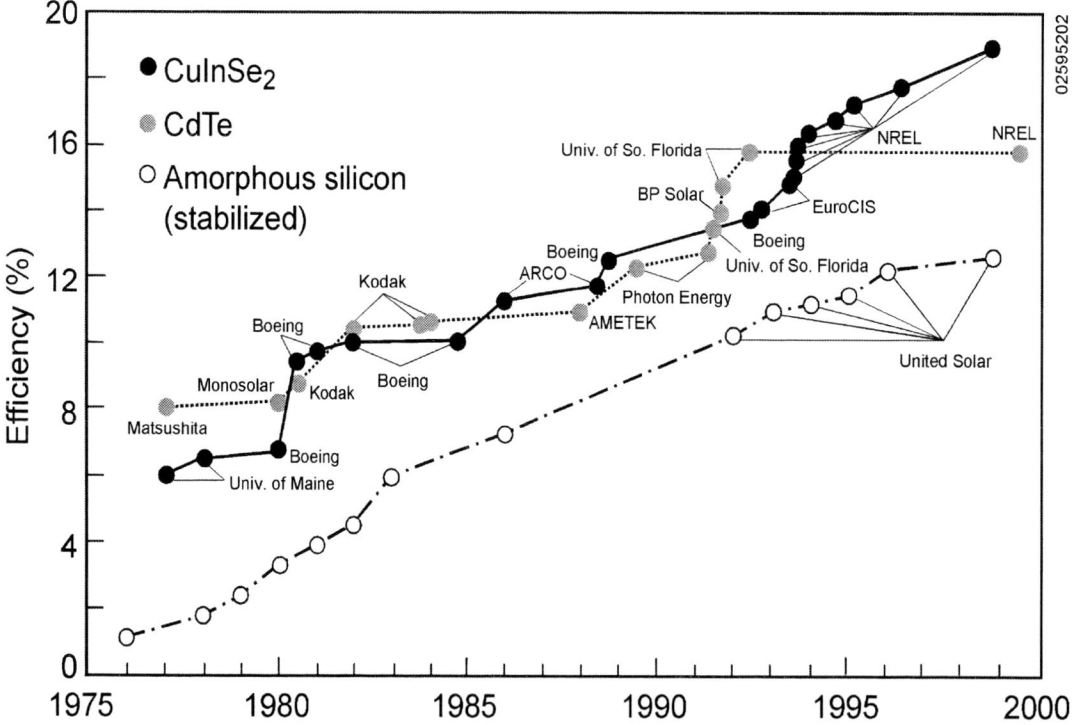

FIGURE 7.3 Thin-film solar cell efficiencies in the laboratory from 1975 until 2000.

a gallium indium phosphide (GaInP$_2$) top cell via a tunnel diode interconnect. The top p/n GaInP$_2$ junction, with a bandgap of 1.83 eV, is designed to absorb the visible portion of the solar spectrum. The bottom p/n GaAs junction, with a 1.42-eV bandgap, absorbs the near-infrared portion of the spectrum, which is transmitted through the top junction. While single gap electrodes have a solar conversion efficiency limit of 32%, tandem junction devices have an efficiency limit of 42%.[6] The maximum theoretical solar-to-electrical efficiency for the present combination of bandgaps is about 34%,[7] and more than 29% efficiency has been realized experimentally. Research work is currently under way to develop systems with four junctions, which have a theoretical efficiency of more than 50% and a realizable efficiency greater than 40%. One approach to deal with the high cost of these materials is to use them in a solar concentrating system, where most of the area of the expensive semiconductor is replaced by an inexpensive optical concentrator. The GaInP$_2$-GaAs system has been shown to operate at up to 1,000 times light concentration (with active cooling). To show the power of this approach, if we take a PV manufacturing plant that is producing 10 MW of PV material per year and the material is capable of 1,000 times concentration, with the use of an optical concentrator system, that plant now produces 10,000 MW (10 GW) of PV per year. While current research involves mainly single-crystal material, applying multijunction technology to thin-film devices would provide great efficiency and cost benefits.

WIND ENERGY

Wind is the world's fastest-growing energy resource. In 2000, worldwide wind-generating capacity was 18,000 MW.[8] (This is equivalent to the amount of nuclear power installed worldwide by 1970.) The growth rate is about 28% per year (1995-1998) and was 36% from 1998 to 1999. Wind systems produce energy at the lowest cost of any renewable energy system, thus wind is projected to produce 10% of the world's energy supply by the year 2020. Wind can be very cost-effective in the displacement of fossil-generated electrons. Electricity from the Lake Benton I 107.25-MW wind farm in Minnesota is sold to Northern States Power Company at an average cost of 3 cents/kWh.

Wind energy turbines range from a few hundred watts to multimegawatt systems. Denmark is currently producing a 2-MW system designed to be used mainly in offshore locations. Because of the lower air turbulence over water, offshore wind turbines produce about 50% more energy than land-based turbines. They also last about 10 years longer.

One of the great advantages of wind is that it is a dual-use technology—its footprint uses only 5% of the land, and the rest of the land can still be used for farming, ranching, and forestry. The land-leasing revenue for a landowner leasing to a wind farm ranges between $500 and $2,000 per turbine per year. That same land used for farming would generate about $100 per year. The farming areas in the Midwest from Texas to North Dakota could be used to provide all of the U.S. electrical needs.

The average capacity factor for wind is about 30%, meaning that the wind farm generates electricity only about 30% of the time. Because of this, utilities often discount wind since they cannot predict or fully depend on the wind farm to generate electricity when it is needed. However, it has been determined that widely separated wind sites provide a more constant power supply; the more wind farms connected to the grid, the more will the short-term fluctuations from one farm cancel out the fluctuations from another. This has been particularly noted in Denmark, which in 2000 was already supplying 13% of its electricity by wind generation. This figure will increase to at least 16% by 2003 and is expected to be 50% of its consumption by the year 2030.

One of the issues with wind is that wind sites are often located away from electrical loads and transmission lines. This also applies to large-scale PV arrays and solar-thermal electric plants situated in the desert areas of the American West.

SOLAR-THERMAL TECHNOLOGY—CONCENTRATING SOLAR POWER

Solar-thermal systems, also known as concentrating solar power plants, produce electrical power by concentrating the sun's energy with various mirror configurations and generating high-temperature heat.[9] They thus follow the path from solar to heat to electricity. The systems are classified by the way in which they concentrate the sun's energy and collect the heat. They include central receiver, parabolic-trough, and dish systems. Although each of these has been demonstrated, the parabolic-trough systems have the longest operating experience.

Parabolic-Trough Technology

Parabolic-trough technology offers the lowest-cost, near-term option for large-scale solar power.[10] It is also the technology that has the longest commercial experience of any solar plant. Electricity costs with current technology are around 10 cent/kWh now, with 5 cent/kWh predicted for the future (2010) using thermal energy storage.

The trough systems collect energy in a receiver pipe located along the focal length of a parabolically curved trough-shaped reflector. A heat-collecting fluid, usually oil, is pumped through the receiver pipes and is heated to about 400°C. The hot oil is then used to produce steam that generates electricity in a conventional steam generator. Systems have been designed that incorporate thermal storage; they can be used to generate electricity when the sun is not shining. However, all parabolic-trough plants currently operate as hybrid solar-fossil plants. These plants use standard fossil fuels to generate electricity when the sun is not shining or during periods of low solar intensity. This points out the great advantage with this technology—it couples nicely with existing fossil fuel technology, thus integrating easily with conventional central station power plants. These hybrid systems provide continuous power, with significantly lower carbon emissions than a stand-alone fossil plant.

There are nine parabolic-trough plants with a combined 354 MW(e) (megawatts of electrical power) of generating capacity operating in the Mojave Desert (Figure 7.4). The first one was installed in 1984. These plants have a combined 100 plant-years of commercial operating experience.

ENERGY STORAGE

Whereas biomass and geothermal can produce energy at will, wind, solar-thermal, and PV cannot. Therefore, any renewable-based energy scheme must have integrated energy storage before it can be considered as a viable, sustainable energy system. Energy storage systems include hydrogen, biofuels, batteries, pumped hydro, compressed air, thermal storage, flywheels, and superconducting magnetic storage. Energy storage is site specific and can be time dependent. Pumped hydro can be very inexpensive but is obviously site dependent. Flywheels and batteries have a high turnaround efficiency (>90%) but degrade at 0.1-5% per hour. Hydrogen has a turnaround efficiency of about 50% but is not time dependent.

Energy must be dispatchable, so only those renewable energy systems that contain a viable energy storage technology can be considered for large-scale implementation. A utility cannot call on solar and wind generation systems to produce at will; it can call on its energy storage system—but the energy must be there! Managing energy storage will be the key for a successful utility. Resource assessment for storage scenarios will be very important for the implementation of renewable energy technologies. The energy storage system will likely be dependent on the local environment, so there has to be the capability to match the energy storage system with the energy generation system. Algorithms and control strategies must be developed to match the generation, storage, and distribution systems. There is a need

FIGURE 7.4 Nine trough power plants in California's Mojave Desert provide the world's largest generating capacity of solar electricity, with a combined output of 354 MW. These systems provide large-scale power generation from the sun and, because of their proven performance, are gaining acceptance in the energy marketplace.

to couple the forecast of energy usage (time and magnitude) with the forecasting of energy generation. Weather forecasts that include wind and solar insolation forecasts become very important. They are lacking at the moment. In fact, the author sees the lack of a large program focused on energy storage and its integration into the energy infrastructure as a major gap in current U.S. energy policy.

Flywheel systems[11] are possibly the lowest-cost energy storage technology, predicted to be in the range of about $500 per kilowatt. Flywheels have a very high turnaround efficiency (>90%), meaning energy in versus energy out, but current technology degrades at about 2% per hour, which limits their energy storage capability to a few hours. Superconducting bearings would give a 0.1% per hour decay[12] and would extend their energy storage capabilities from days to weeks. Only kilowatt-hour energy storage systems have been demonstrated. For utility applications, multimegawatt-hour systems will have to be developed and demonstrated.

Hydrogen can replace fossil fuels as the energy carrier for transportation and electrical generation when renewable energy is not available. Since it can be transported via gas pipelines or generated on-site, any system that requires an energy carrier can use hydrogen. Conversion of the chemical energy of H_2 to electrical energy via a fuel cell produces only water as waste. This pollution-free attribute of a hydrogen economy is one of the major driving forces for research in the generation, storage, distribution, and utilization of hydrogen.[13]

Currently, hydrogen is manufactured in large quantities from steam reforming of natural gas; however, H_2 from solar energy can be generated through a number of paths.[14] The conversion of biomass to

H_2, although fairly straightforward, has a low sunlight-to-hydrogen conversion efficiency, and any system designed to generate significant amounts of H_2 must cover a rather large land area. Nonetheless, if the biomass is a waste by-product (such as wood chips or nutshells), then this is perhaps the least expensive of the renewable H_2 generation technologies. Wind energy and photovoltaic systems coupled to electrolyzers are perhaps the most versatile of the approaches and are likely to be the major hydrogen producers of the future. These systems are commercially available, but only at high prices. Advanced photolysis systems combine the two separate steps of electrical generation and electrolysis into a single system. These direct conversion systems include photoelectrolysis and photobiological systems. These systems are based on the fact that visible light has sufficient energy to split water.[15]

The coupling of hydrogen to the energy and transportation infrastructures requires the development and deployment of fuel cell technology. Fuel cells are devices that take chemical energy and, without combustion, convert the fuel directly to electricity.[16] Since there is no burning of the fuel, there is no generation of airborne pollutants; the only effluent is water. High-efficiency fuel cell technologies are well known; perhaps the best-known current application is the use of fuel cells to power the space shuttle. Fuel cells represent the most energy-efficient link between renewable-based fuels (hydrogen, methanol, ethanol, and other biofuels) and electricity.

In a typical fuel cell, hydrogen and oxygen are supplied to individual electrodes separated by an ion-conducting layer. The electrochemical reactions produce electricity, heat, and water. Fuel cells are classified by the electrolyte used. These include the proton exchange membrane fuel cell (PEMFC), phosphoric acid fuel cell (PAFC), alkali fuel cell (AFC), molten carbonate fuel cell (MCFC), and solid oxide fuel cell (SOFC). Solid-electrolyte fuel cells such as the SOFC and PEMFC have a number of advantages in comparison to other types. The solid electrolyte removes liquid electrolyte management problems, reducing or eliminating corrosion, leakage, pore flooding, and so forth. The high temperature used in the SOFC system (>650°C) promotes rapid kinetics and eliminates the need for precious metal catalyst, but places severe requirements on materials, as well as requiring thermal management. It also dictates an extended startup and shutdown period, virtually eliminating its use for intermittent power generation. PEM systems operate at lower temperatures (<90°C), but they require strict water management for membrane humidification and platinum or other precious metal catalysts to increase the kinetics.

The strategic potential of fuel cell technology is enormous. Fuel cells integrate into renewable energy power packages and facilitate distributed generation. Biomass-fired fuel cells, where gasified biomass is fed into molten carbonate or solid-oxide fuel cells, produce a concentrated CO_2 stream that if sequestered, actually reduces atmospheric CO_2. Fuel cells can be integrated into building energy systems to provide both heat and electricity. Fuel cells are the key enabling technology for future transportation vehicles, providing high efficiency, high fuel economy (80-100 mpg [miles per gallon] equivalent), and zero emissions.[17]

Fuel Cell Conundrum—The Current Fuel Infrastructure Won't Work

For fuel cell vehicles to be acceptable to consumers there must be a readily accessible and reasonably priced fuel for refueling. However, a production and distribution system for a new fuel will not be built unless there is a known demand. The most versatile way to solve this problem is with electrolysis. Electrolysis uses electricity in an electrolyzer to break apart water into hydrogen and oxygen. Although current commercial electrolyzers are very expensive and used primarily in industry to produce high-purity H_2, companies are working to bring the costs down and develop small inexpensive systems suitable for individual use.

The approach would be to provide a "home fueling appliance" (i.e., an electrolyzer) with each fuel cell vehicle.[18] The cost goal is about $1,500 per appliance and would be included in the vehicle cost. The consumer plugs the appliance into an outlet in the garage, and water is delivered via a garden hose. There would be no storage of hydrogen in the appliance; the consumer would connect the appliance up to the vehicle, which would receive its charge of hydrogen overnight. It should be recognized that with this approach, the consumer is purchasing most of her or his fuel up-front, with reoccurring charges from the electricity used to power the electrolyzer.

Electrolyzers in the Infrastructure

Using electrolyzers to provide hydrogen for fuel cell vehicles with our current fossil-based energy infrastructure actually doubles the CO_2 produced per mile compared to commercial internal combustion engine technology[19] because of the low efficiency of the current electricity generation system and the efficiency of electrolysis. Therefore, this approach is viable only if there is a concurrent program for major introduction of renewable generation of electricity.

Alternatively, small-scale reformers (100-1,000 cars per day) generating hydrogen from natural gas would cut emissions CO_2 per mile by half. This approach may be viable for the short term, but it is not sustainable in the long term.

Water Issues

Water is already an issue for current fossil-fueled plant construction, especially in the arid western United States. Water is also a global issue.[20] It is apparent that generating hydrogen from water electrolysis will only exacerbate the issue. The conclusion is that water desalination plants will be necessary. If these desalination plants are placed in or near cities that currently have oil ports, then the existing liquid fuels distribution infrastructure could be used for "electrolysis-grade" water delivery.

PATHWAY TO A RENEWABLE ENERGY INFRASTRUCTURE

What would be the most reasonable approach to replacing our current energy infrastructure with a sustainable one? The author respectfully offers the following recommendations.

Any renewable energy system that produces electrons should be connected to the grid in such a way as to directly reduce the CO_2 emissions from current fossil energy generation and avoid construction of additional fossil fuel power plants.[19] PV comes very close to matching the early afternoon peak in energy use. Integrating PV into current buildings and future building designs should be strongly encouraged. Wind farms, being the least-cost renewable energy systems, should be encouraged with incentives and legislative mandates. There is some debate as to how much intermittent power the grid can accept, but even 10% would constitute a large amount of renewable energy and greatly reduce CO_2 generation.

Biomass-based power plants and parabolic-trough hybrid systems could integrate seamlessly into the current infrastructure, providing continuous power and lowered CO_2 emissions.

Fuel cell vehicles should be deployed with hydrogen as the on-board fuel. Hydrogen is the preferred fuel because it provides the greatest benefit in terms of fuel economy and emissions.[19]

During the buildup of the renewable generation systems, research and development should be focused on energy storage technologies in the multimegawatt-hour to gigawatt-hour range. These would be deployed as the technology is proven.

Only minimal sequestration should be considered because it costs money and energy, and the money and energy are better spent deploying a sustainable energy infrastructure.

RESEARCH AND DEVELOPMENT ISSUES

Because renewable energy technologies must cover an area for the collection of energy, the efficiency of every process is important. Efficiency also relates directly to costs. These efficiencies include the following:

- Efficiency of electrical generation (PV, wind, solar thermal, etc.)
- Efficiency of energy storage (hydrogen, flywheel, etc.)
  1. For hydrogen: electrolysis efficiency and fuel cell efficiency
  2. For flywheel: bearing losses
- Efficiency of utilization
- Efficiency of system coupling

Silicon Solar Cell Technology

The PV industry has been very good at reducing costs; however, it is going to run up against a barrier in the cost of the silicon feedstock used to make solar cells. Low-cost, low-energy technologies must be developed that can take the raw material (quartz) and refine it into solar-grade silicon.

Techniques are being developed to grow thin layers of silicon on various substrates to minimize the amount of silicon used in the manufacture of solar cells. Research is needed to determine ways to make thin-film silicon perform at high efficiencies and, in particular, how to mitigate the effect of grain boundaries.

Thin-Film Solar Cell Technology

For production in the range of about 30 GW per year the material availability with current technologies—particularly for the elements indium, gallium, tellurium, and germanium—would cause supply issues. Research is needed for the discovery and development of new thin-film semiconductors that will reduce or eliminate the necessary amount of these materials.

Research is needed for the discovery and development of new thin-film semiconductors that will replace current toxic and heavy metals (cadmium, tellurium, lead) with nontoxic materials. In the meantime, in case they cannot be eliminated, research is needed on the recovery and recycling of these toxic materials.

Work is needed on the replacement of toxic or explosive feedstock gases that are used in the manufacture of thin-film systems. Multijunction thin-film systems have to be developed for increased solar-to-electrical conversion efficiency. Continued work is necessary on the fundamental mechanism for the degradation of amorphous silicon devices. There may also be some stability issues with the other thin-film technologies.

Wind

Technical issues for improved turbine performance and lower costs include aerodynamics, structures and fatigue, advanced components, and wind characteristics.

Basic aerodynamics research for three-dimensional computer simulations of airflows is rarely used in the aircraft industry, so wind turbine researchers have to develop new methods and computer simulation models to deal with these issues. Research in computational fluid dynamics (CFD), which is a group of methods that deal with simulating airflows around, for example, rotor blades for wind turbines, is also needed.

Materials engineering is needed for advanced components to improve performance and reduce hardware costs. Research into innovative generators and advanced controls, including power electronics, is needed.

Other activities that must be conducted include developing an updated, comprehensive national database for utility and industry access and improving resource assessment and mapping techniques and wind forecasting.

Solar-Thermal

Research issues related to increasing the efficiency and decreasing the costs of solar-thermal technology include the following:

- Optical materials—durability, flexibility (easily applied to compound curvature surfaces), high reflectivity, easy cleanability, low cost
- Concentrators (heliostats and dishes)—low-cost drives, lightweight structures, high optical accuracy, flexible control systems, low-cost, innovative system concepts.
- Receivers—high-efficiency volumetric reactors, secondary concentrators
- Storage—high-temperature, low-cost storage concepts

Electrolyzers and Fuel Cells—PEM Systems

One cannot emphasize enough the necessity for increased research in electrocatalysts. Hydrogen will be one of the main components of a renewable energy infrastructure, and the major conversion technologies for hydrogen, namely electrolyzers and fuel cells, both involve electrocatalysts. Catalysts will be needed for the oxygen reaction for both water oxidation and oxygen reduction. Other research that is needed for electrolyzers and PEM fuel cells includes the following:

- More reactive catalysts for direct methanol (and perhaps ethanol) fuel cells must be developed.
- Ways are needed to reduce (or better substitute for) platinum (180 g per system now for a 50-kW(e) fuel cell).
- Better manufacturability of fuel cells is necessary; including.
    1. better manufacture of bipolar plates (injection molded or stamped from metal stock) and,
    2. better membranea (Nafion—fluorine chemistry).

REFERENCES

1. To estimate the land area needed for photovoltaic panels, the following information was used (from R.L. Hulstrom, National Renewable Energy Laboratory internal report): Flatplate photovoltaic (PV) collector modules are typically placed such that they cover one-half of the available land; 1 m^2 of PV requires 2 m^2 of available land. Total U.S. 1997 annual electricity demand was about 3.2×10^{12} kWh. Average solar resource per year for southwest Nevada is 2,300

kWh/m^2 per year. If we assume 10% net plant efficiency (current technology), that solar resource would provide 230 kWh/m^2 per year. Therefore the total area needed is

$$\frac{3.2 \times 10^{12} \text{ kWh/y}}{230 \text{ kWh/m}^2 - \text{y}} = 1.39 \times 10^{10} \text{ m}^2 \text{ of collector area,}$$

which equals 2.78×10^{10} m^2 of land area, or about 10,900 square miles (a square 104 miles on one side). A system efficiency of 15% would drop the area to 7,200 square miles.

2. Alsema, E.A., and E. Nieuwlaar. *Energy Policy* 28(14):999-1010 (2000).
3. Oliver, M., and T. Jackson. *Energy Policy* 28(14):1011-1021 (2000).
4. See http://www.windpower.dk/tour/env/enpaybk.htm.
5. See http://www.nrel.gov/ncpv/.
6. Bolton, J.R., S.J. Strickler, and J. S. Connolly. *Nature* 316:495 (1985).
7. Kurtz, S.R., P. Faine, and J. M. Olson. *J. Appl. Phys.* 68:1890 (1990).
8. See http://www.awea.org/faq/ and http://www.windpower.dk/core.htm.
9. See http://www.eren.doe.gov/csp/.
10. See http://www.eren.doe.gov/power/pdfs/solar_trough.pdf and H. Price and D. Kearney. A Pathway for Sustained Commercial Development and Deployment of Parabolic-Trough Technology, NICH Report No. TP-550-24748 (1999), found at http://www.nrel.gov/docs/fy99osti/24748.pdf
11. Hull, J.R., *IEEE Spectrum* 34(7):20-25, (1997).
12. Hull, J.R., *Superconductor Science & Technology* 13(2):R1-15 (2000).
13. See http://www.eren.doe.gov/hydrogen.
14. Turner, J., *Science* 285:629-792 (1999).
15. Khaselev, O., and J. Turner. *Science* 280:425 (1998).
16. See http://education.lanl.gov/resources/fuelcells.
17. Thomas, C.E., B.D. James, and F.D. Lomax, Jr. Analysis of Residential Fuel Cell Systems and PNGV Fuel Cell Vehicles, in Proceedings of the 2000 Hydrogen Program Annual Review, found at http://www.eren.doe.gov/hydrogen/pdfs/28890mm.pdf.
18. Fairlie, M.J., and P.B. Scott. Filling up with Hydrogen 2000, in Proceedings of the 2000 Hydrogen Program Annual Review, found at http://www.eren.doe.gov/hydrogen/pdfs/28890z.pdf.
19. Thomas, C., B. James, F. Lomax, and I. Kuhn. Integrated Analysis of Hydrogen Passenger Vehicle Transportation Pathways, in Proceedings of the 1998 Hydrogen Program Annual Review, found at http://www.eren.doe.gov/hydrogen/pdfs/25315o.pdf.
20. Vorosmarty, C. J., P. Green, J. Salisbury, and R. Lammers. *Science* 289:284 (2000).

APPENDIX A

The following is an example of a residential distributed energy system utilizing renewable generation and an electrolyzer-hydrogen storage-fuel cell combination for energy storage.

The energy system for this home utilizes renewable energy generation (Solar cells on the roof of the home) and hydrogen storage to provide continuous power. (See Figure 7.5.) The key to this house is the controller. The controller is connected to the Internet and monitors and controls all of the components. The components of this system include an electrolyzer that utilizes electricity to split water into hydrogen and oxygen, a hydrogen energy storage system, and a fuel cell to convert the stored hydrogen back into electricity. Using parameters set by the homeowner (similar to a thermostat), the controller maintains the state of charge for the hydrogen system and either purchases or sells power to the grid. If the sun is shining on the PV panels, electricity can be either used in the home, directed toward the electrolysis unit, or sold to the grid. At night, hydrogen is supplied to the fuel cell to generate electricity. Alternatively, if available, the controller could purchase off-peak power from the grid to operate the home and electrolyze water to generate additional hydrogen. The hydrogen would be stored in underground tanks, either as a pressurized gas, absorbed into advanced carbon nanotubes, or as a metal

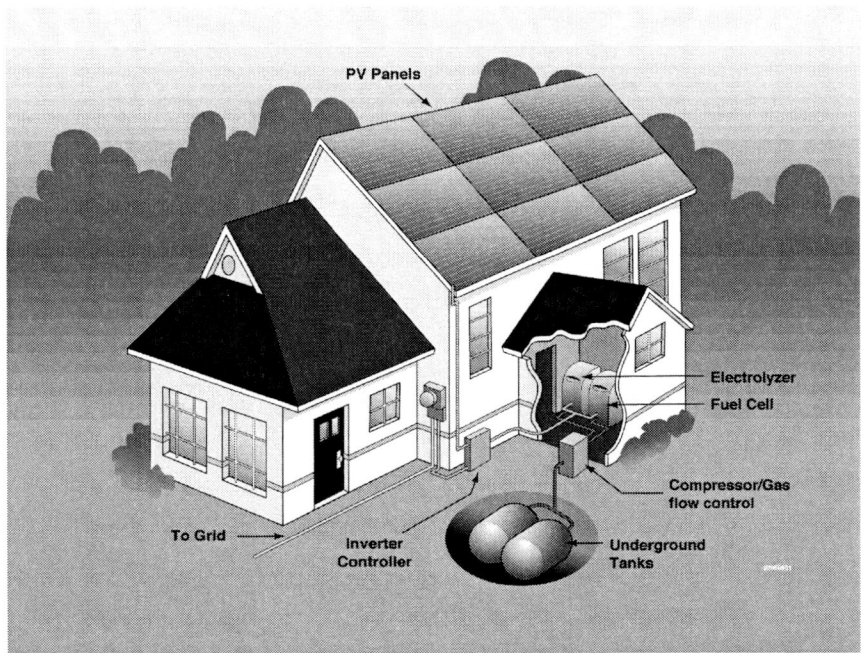

FIGURE 7.5 Examples of a residential distributed energy system utilizing renewable generation and an electrolyzed/hydrogen storage/fuel cell combination for energy storage.

hydride. In addition to providing energy storage, the hydrogen could also be used to fuel the homeowner's fuel cell vehicles.

As an option, the homeowner could make power available to the electric utility. If the utility has a need for additional power, it would send out a query to the controller to negotiate a price and to determine the amount of power that would be available from the home. Power could come from either stored hydrogen via the fuel cell or PV electricity. While each home would not necessarily have a lot of power to sell, a large number of homes (a hundred thousand or more) would represent a very large energy storage and generation system. Communities of homes could act as energy generation and storage systems for local industry, the ultimate in distributed generation.

DISCUSSION

Dave Cole, Oak Ridge National Laboratory: I have two questions that pertain to silicon. First, on the issue of stability, could you elaborate on what you really mean by stability of the silicon-based compounds?

Second, with regard to the thin film, at what degree of degradation of the thin film does it become inefficient or does the efficiency drop off so much that it can't function anymore?

John Turner: In terms of silicon, there are two issues and I zipped through it so quickly that it has caused confusion. There is no issue of physical degradation of crystalline silicon; as long as you keep it dry, it lasts 30 years. The degradation is in the performance, and that has a stable, lasting effect. It is not a really well understood effect, but under light soaking, the system loses some of its efficiency in the first six months of operation.

So, the films start out at 15% and then in six months they drop down to 10 or 12%. This particular degradation mechanism is somewhat understood, but how to control it is not.

Dave Cole: Is moisture an issue in temperate zones where it is not always dry? Is water or humidity a factor?

John Turner: Water is not an issue with current sealing technologies for single-crystal silicon cells.

Dave Cole: Is that true for the amorphous silica also?

John Turner: It is the same for amorphous silica. Amorphous silicon uses the same sealing technique.

Dave Cole: What do you mean by "sealing technique"?

John Turner: Typically, the silicon is deposited either on glass or on stainless steel and sealed with a plastic. The plastic is called ethoxybenzoic acid (EBA), but I've forgotten what that means now. It is a cover that seals everything up. It lasts about 20 years. The degradation of EBA is really what limits the lifetime of silicon solar cells.

David Keith, Carnegie Mellon: I think I heard you say that the use of water in electrolysis to make hydrogen would lead to some water shortages. Is that correct?

John Turner: It is going to be an issue if you only have water—

David Keith: I don't get it. I mean about a meter of water a year falls in the typical temperate zones. That means on each hectare you get 10^4 square cubic meters of water per year.

John Turner: Yes.

David Keith: Yet the amount of water that you need to make hydrogen, assuming you—

John Turner: Pretty small.

David Keith: It is of order 10 cubic meters. So, there is a difference of 10^3.

John Turner: Right, but it is the culmination that makes the difference. Water is used for food, people and industry. If you add hydrogen production to that, there may be a problem. If you take it by itself, you are absolutely right.

David Keith: It is down by three orders of magnitude.

John Turner: It is going to be an issue.

8

Industrial Carbon Management: An Overview

David W. Keith
Carnegie Mellon University

It is possible to use fossil fuels without atmospheric emissions of CO_2. I call the required technologies industrial carbon management (ICM)—defined as the linked processes of capturing the carbon content of fossil fuels while generating carbon-free energy products, such as electricity and hydrogen, and sequestering the resulting carbon dioxide. Although many of the component technologies currently exist at large scale, the idea that ICM could play a central role in our energy future is a radical break with recent thinking about energy systems and the climate problem.

This chapter aims at a synoptic view. I first introduce the core technologies required to implement ICM and describe their roots in the existing fossil fuel infrastructure. I then speculate about how these technologies might diffuse into the current infrastructures for energy distribution and use and about how the existence of viable ICM technologies might affect the overall cost of mitigating CO_2 emissions. Finally, I consider the implications of ICM, first for the management of R&D in the chemical sciences and then for the broader politics of the CO_2-climate problem.

THE CO_2-CLIMATE PROBLEM[1]

The CO_2-climate problem is not new. Although for many of us this topic appeared only yesterday, you may judge its age by reading a few of the beautifully written reports of the 1960s and 1970s. *Restoring the Quality of Our Environment* (President's Science Advisory Committee, 1965) has a chapter on CO_2 and climate that demonstrates the problem in its modern form. The report first analyzes the growth of atmospheric CO_2 by comparing the Keeling record of accurate concentration measure-

---

[1]This essay is loosely based on a presentation at the National Academy of Sciences' *Chemical Sciences Roundtable*. In transforming it to published form, I have attempted to preserve some of the informality of a verbal presentation while (I hope) improving its organization and content. There was considerable skepticism expressed during the meeting about the scientific basis for concern about climate. I tried to answer this skepticism with brief opening comments; I have included them here, although they differ in style and content from the rest of this essay.

ments (initiated in 1958) with estimates of global fossil fuel combustion and then makes crude estimates of future concentrations. It then combines concentration estimates with early radioactive connective models to estimate temperature change, and compares estimated changes to observed changes with consideration given to intrinsic climate variability. Finally, it speculates about possible impacts beyond temperature such as the CO_2 fertilization of plant growth.

The National Academy of Sciences 1977 report on *Energy and Climate* serves as another demonstration of the consistency in our understanding of the climatic implications of transferring large quantities of carbon from fossil reservoirs to the atmosphere. When compared to recent Intergovernmental Panel on Climate Change (IPCC) reports, these older studies show that although there is a wealth of new science, what is really new is the growing will toward real action—and sadly an unambiguous trend to toward increasing in the length and opacity of climate assessment reports.

Are we seeing an anthropogenic climate signal yet? What signal will we see if we double or triple CO_2 concentrations? These are the two crucial scientific questions. Although they are often deliberately blurred in public debate, they are sharply distinct. On the first, my answer is we cannot yet claim to have seen an unambiguous anthropogenic signal. Although we can measure the signal (the climate record) reasonably well, the crux of this so-called detection and attribution problem is our uncertainty about the climate's unforced variability over decade-to-century time scales—the climate's noise spectrum—and our uncertainty about the magnitude of natural forcing such as solar variability. Without robust understanding of the noise we can't calculate the signal-to-noise ratio very confidently and thus cannot yet answer the detection question unequivocally.

It is the second question, however, that ought to matter for policy: If we double or triple CO_2, will we see a big signal? Here the answer is an unequivocal yes. With the exception of a few outspoken individuals, there is essentially no serious scientific dissent on this question.

When you move beyond the climate science to consider the impacts of climate change, the answer bifurcates again. If you ask, "will there be substantial impacts on many lightly managed ecosystems and on some human populations?" the answer is unequivocally yes. Examples include the ecosystems of the high Arctic and the inhabitants of low-lying islands.

If, however, you want to know about the net economic impacts of climate change, the answer is much less certain. Most economic modeling studies suggest that if we do nothing to abate climate change, we will suffer a loss of a few percent of global gross domestic product (GDP) by 2100 and that modest efforts at abatement are capable of reducing this loss (e.g., the benefits of abatement can outweigh the costs). Modeling our economic future on 100-year time scales is, however, an uncertain and untested art; both the costs of mitigation and the estimates of impact are highly uncertain.

When judged in gross economic terms, climate change will not be catastrophic and is in fact (arguably) minor compared to other economic uncertainties such as the rate of technological change or the evolution of the inequality in distribution of the world's wealth. Despite some overheated rhetoric, our civilization will not collapse if we fail to act aggressively to counter climate change. Humans are very robust. Nevertheless, climate change poses very serious environmental problems. If we double or triple carbon dioxide—as many business-as-usual projections suggest we will within this century—we will see substantial climate change and very substantial changes to many natural ecosystems. Many of us value these ecosystems and value the rights of human communities that will be most affected. We have shown by other actions that we are willing to pay a price to protect such values. I hope and expect that we will take strong action to stabilize CO_2 concentrations. The history of pollution control technologies gives reason to hope that the price of achieving deep cuts in CO_2 emissions will be lower than we predict. The topic of this meeting gives us an explicit basis for that hope.

INDUSTRIAL CARBON MANAGEMENT

Use of fossil fuels with minimal emissions of CO_2 requires two steps. The energy content of the fuels must first be *separated* from their carbon content in a process that takes a fuel with high carbon-to-energy ratio as input and produces a low-or zero-carbon output along with a carbonaceous stream with low free energy to be *sequestered* away from the atmosphere. In practice, this generally means a system that uses coal or gas to produce electricity or hydrogen with sequestration of the resulting CO_2, but other options may prove important (Figure 8.1). Separation and sequestration together comprise industrial carbon management.

These technologies are not laboratory theory; on the contrary, many of the required components exist at the largest industrial scales. Among the most important such technologies are the gasification of coal, the capture of CO_2 using aqueous amines, the steam reforming of methane, and finally, the long-range transport of CO_2 by pipeline and its injection into geological formations. We have, in essence, a sizable basket of component parts out of which we might assemble a system for carbon management.

Although ICM makes sense only as part of a broad portfolio of greenhouse gas mitigation technologies, it may nevertheless transform the politics of the CO_2-climate problem. By lowering the cost of emissions mitigation, ICM may enable stabilization of atmospheric concentrations at acceptable cost. By weakening the link between fossil energy and atmospheric CO_2 emissions, ICM makes it feasible to consider a fossil-based global economy through the next century. By reducing the severity of the threat that emission reduction poses to fossil industries and fossil-rich nations, ICM may ease current deadlocks in both domestic and international abatement policy.

There are, however, no magic bullets with which to slay the CO_2-climate problem. All current options—ICM included—either are impracticably expensive or involve significant environmental risks. Moreover, global energy systems are highly heterogeneous, making it implausible that any single technology will triumph everywhere. Finally, the history of energy policy is full of technologies that seemed to their advocates to be too cheap to meter, yet are now irrelevant. Thus, although I will tell you a (mostly) optimistic story about the potential role of ICM in mitigating CO_2 emissions, skepticism is

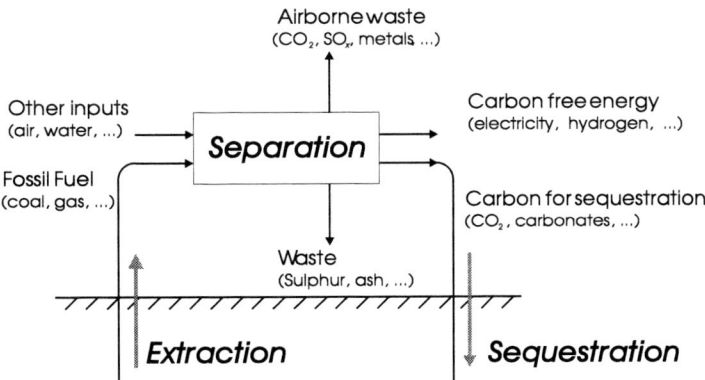

FIGURE 8.1 Industrial carbon management: a schematic illustrating the definitions of *sequestration* and *separation* adopted here. The output stream is labeled *carbon free* for simplicity; *separation* includes processes that take a high-carbon stream in and produce a lower carbon-to-energy product (e.g., coal to natural gas with sequestration). All methods of ICM involve an energy penalty—the output stream contains less energy than the inputs.

wise. The very fact that ICM was not on the energy policy agenda even a decade ago should make one cautious about any predictions for the next century.

SEPARATION

There are three broad paths to separation:[2]

1. *Postcombustion capture* (PCC): Burn the fuel in air then capture CO_2 from the combustion products.
2. *Oxyfuel*. Separate oxygen from air, burn the fuel in pure oxygen, and then capture CO_2 from the combustion products.
3. *Precombustion decarbonization.* (PCDC). Reform fuel to make hydrogen and CO_2.

The first two, PCC and oxyfuel, involve complete combustion to CO_2 and water and so are limited to producing electricity and heat, whereas PCDC produces hydrogen that may be combusted in an integrated system to produce electricity and heat or may be distributed for use elsewhere.

Most discussion of ICM has focused on large-scale electricity generation, where PCC is perhaps the most obvious route to separation because it is closely analogous to existing environmental control technologies, such as flue gas desulfurization, that remove pollutants from power plant exhaust streams. Amine solvents are now used to capture CO_2 from power plant exhaust streams for commercial uses such as the carbonation of beverages. Using current technology, amine systems are able to capture about 90% of the CO_2, but the energy cost of solvent regeneration reduces plant electrical output by about 15%. A host of other capture methods have been proposed, and there is evidence that amine technologies can be significantly improved. In comparison to other routes to separation, PCC has the great advantage that it requires little modification to existing power plants and so could in principle be applied as a retrofit; its disadvantage is that the separation is performed at atmospheric pressure on a gas stream that is only 4 to 15% CO_2 and contains a multitude of reactive combustion products. From a coal-fired plant, for example, the exhaust gas is at most about 15% CO_2 and contains SO_x, NO_x, and various metals so that a CO_2 separation system must either be tolerant of impurities or the impurities must be removed. Significant opportunities exist for co-optimization of the multiple emissions control technologies that must be applied to coal-fired plants.

Instead of separating CO_2 from the combustion gases, you can first separate the O_2 from air and then do the combustion in pure O_2, producing an exhaust stream that is CO_2, H_2O, and impurities from which the water can easily be removed by condensation. Compared to PCC, Oxyfuel schemes offer the advantage of doing the primary separation step on a clean gas mixture (air) that is free from the many reactive impurities in combustion gases. The leading disadvantage of Oxyfuel is the high energy and capital cost associated with oxygen production. Oxyfuel systems also offer higher combustion temperatures that yield higher Carrot efficiencies. The flame temperature from pure Oxyfuel is too high, however, so all Oxyfuel schemes must use a diluent to reduce the temperature and increase the volume of working fluid. The leading choices are direct injection of water or recycle CO_2. The direct (i.e., without reforming to produce H_2) use of methane in high-temperature fuel cells may be considered an Oxyfuel route to separation because an oxygen-permeable membrane is used to transport O_2 to CH_4 for oxidation. This analogy is particularly apt because one of the leading methods to produce O_2 in combus-

---

[2]The process is often called CO_2 capture rather than separation, but I prefer separation because not all processes end with CO_2. It is also possible to produce carbonates or even pure carbon.

tion-based Oxyfuel schemes involves the use of air separation membranes that are closely related to the membranes used in solid oxide fuel cells.

PCDC is most obviously accomplished by steam reforming of methane followed by the water-gas shift reaction to produce a CO_2-H_2 mixture. Separation of CO_2 from such gas streams is much easier than it is from combustion air in PCC systems because of the higher working pressures and higher fraction of CO_2. As described below, this method is now used to produce hydrogen from methane at very large scale. Many other methods are possible. For methane, one can, for example, produce synthetic gas using partial oxidation instead of steam reforming. For coal, a multitude of gasification reactions are possible including, for example, H_2 rather than O_2-blown gasification. Even in a electric power plant, PCDC systems appear to be competitive with other methods whether the fuel is coal or natural gas. In addition, PCDC has the important advantage that a power plant could sell zero-CO_2-emission hydrogen "over the fence" to support the development of a hydrogen infrastructure (Ogden, 1999).

GEOLOGICAL SEQUESTRATION[3]

Although much is uncertain about geological sequestration, the essence of current knowledge is easily stated: (1) it is possible to put very large volumes of CO_2 underground at comparatively low cost; (2) it appears that a capacity of greater than 1,000 gigatonnes carbon (GtC) exists in reasonably well understood geological structures; and, (3) while the fate of CO_2 is highly dependent on the specific geological character of the injection site, it seems highly likely that a large fraction of CO_2 could be confined underground for time scales in excess of a thousand years.

As we now envision it, the CO_2 would be injected into geological formations similar or identical to the formations from which we now extract oil and gas, and the technologies employed would be readily derived from current systems used in the oil and gas industry. The most likely sequestration sites and their estimated capacities are shown in Figure 8.2.

While geological sequestration will build generally on the totality of experience with fossil fuel extraction, it will be most directly built on current practice of CO_2 injection for enhanced oil recovery (EOR). Conventional extraction methods typically leave substantial oil in place. This oil may be extracted using EOR. Carbon dioxide injection (or "flooding" in industry jargon) is particularly effective because, as an organic solvent, the CO_2 acts to reduce the viscosity of the residual oil and in addition causes the oil to expand thus helping to free it from the porous rock in which it is embedded. Typical EOR floods operate at pressures above the critical point of CO_2 so that fluid flow is facilitated by the absence of a liquid-gas interface.

The use of CO_2 for EOR will provide early sequestration opportunities at negative cost as EOR operations pay of order $50/tC for CO_2. Most assessments suggest that absent EOR, the cost of geological sequestration will be of order $10-25/tC. At this price, the overall cost of ICM is dominated by the cost of separation. There are reasons, however, to expect that these estimates may be too optimistic and that sequestration cost will consume a rising fraction of the total cost of ICM. Costs of CO_2 sequestration may be higher than predicted from the EOR experience due to the additional costs involved in monitoring and verification. Depending on our experience with CO_2 injection and on the regulatory framework that is adopted (which will likely be different for current regulation of CO_2 EOR) the costs of monitoring could be very high. Moreover, sequestration cost estimates have tended to assume injection into previously characterized high-permeability structures, but in the long run these will be satu-

---

[3]Oceanic sequestration is possible, and perhaps important, but I omit discussion of it here.

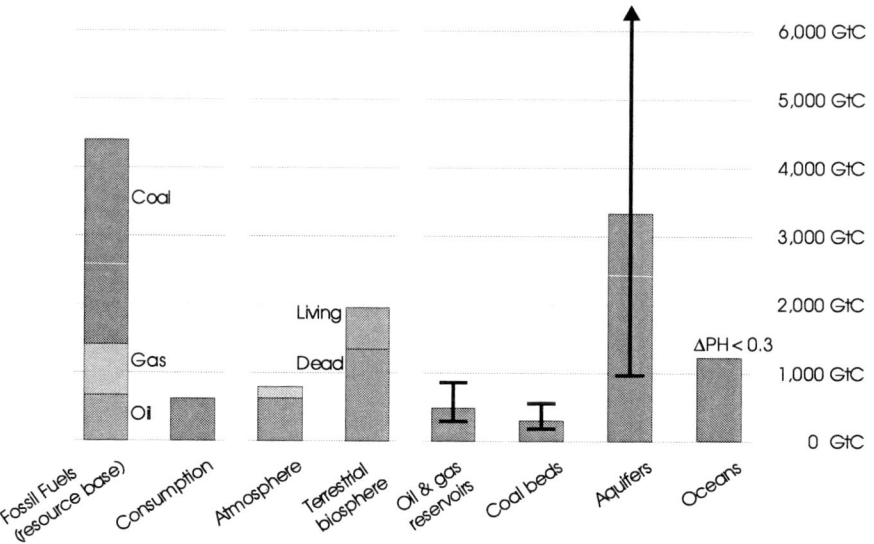

FIGURE 8.2 Carbon reservoirs and sinks. The resource base (the sum of reserves and resources) is used for fossil fuels [Rogner, 1997]. The *consumption* box shows worldwide cumulative consumption of fossil fuels. The upper section of the *atmosphere* box shows the increase in CO_2 since preindustrial times. The error bars are a rough summary of current knowledge and do not reflect systematic analysis of uncertainty. The upper bound for storage in aquifers is of the order 10,000 GtC. The oceanic capacity is based on an arbitrary upper limit to pH change of 0.3; surface ocean pH has already decreased by ~0.1 due to anthropogenic CO_2.

rated and we will have to turn to lower permeability structures and to include the full cost of subsurface characterization. Arguably this could drive sequestration costs closer to the cost of natural gas extraction (currently of the order $100/tC).

Carbon need not be sequestered as CO_2; instead, it can be sequestered as a stable and immobile carbonate. The process mimics the natural weathering of magnesium and calcium silicates that ultimately react to form carbonate deposits. Integrated power plant designs have been proposed, in which a fossil fuel input would be converted to carbon-free power (electricity or hydrogen), with simultaneous reaction of the CO_2 with serpentine rock (magnesium silicate) to form carbonates (Lackner et al., 1995). Carbonate formation is exothermic; thus, in principle, the reaction requires no input energy. Ample reserves of the required serpentine rocks exist at high purity. The size of the mining activities required to extract the serpentine rock and dispose of the carbonate are comparable to the mining activity needed to extract the corresponding quantity of coal. The difficulty is in devising an inexpensive and environmentally sound industrial process to perform the reaction.

The importance of geochemical sequestration lies in the permanence with which it removes CO_2 from the biosphere. Unlike carbon that is sequestered in organic matter or as CO_2 in geological formations, once carbonate is formed the only important route for it to return to active biogeochemical cycling is by thermal dissociation following the subjection of the carbonate-laden oceanic crust beneath the continents, a process with a time scale of more than 10^7 years.

Risks

The risks of large-scale underground sequestration of CO_2 are poorly understood, and there has been little or no systematic effort at risk assessment. The risks may be roughly divided into two kinds: (1) the direct risks to humans and local environments, and (2) the risk of slow leaks that return the sequestered carbon to the atmosphere.

The most obvious of the direct risks is due to catastrophic release of CO_2, but there are also hazards from slow leaks and possible risks stemming from underground movement of displaced fluids such as induced seismically or contamination of potable aquifers. Experts in the upstream oil and gas industry are generally confident that the risks are small, and this confidence is strongly supported by the long history of CO_2 injection for EOR and of underground storage of other gases, including the very large scale storage of natural gas. Nevertheless, the basis for concern is clear. Natural gas storage facilities have leaked to the surface causing dangerous buildup of gas in buildings, and natural emissions of CO_2 can pose serious risks. In 1986, for example, Lake Nyos (Cameroon) released a dense CO_2-rich cloud that killed more than 1,700 people by suffocation. Here in the United States, widespread deaths of trees and one possible human fatality in the last decade have been linked to degassing of CO_2 from the Long Valley Caldera in the Mammoth Lakes area, California. A very recent death (July 2000) in a naturally occurring soda springs bath at Clear Lake, California, underlines the constant danger posed by CO_2 emissions from the ground.

All separation technologies extract an energy penalty so that more fuel must be consumed—and more CO_2 produced—per unit of delivered energy than would be the case if the CO_2 were not captured. In the worst case, therefore, if the CO_2 is rapidly returned to the atmosphere, ICM can *increase* future concentrations of CO_2. Simple modeling of underground transport suggests that lifetimes in excess of 1,000 years can readily be achieved, and evidence from natural CO_2 formations suggests that retention times can be orders of magnitude longer. While there is ample reason to expect that sufficiently low leak rates can be achieved, it is not yet possible to specify with confidence the site characteristics and injection technology required to ensure (within a defined level of uncertainty) that a given leak rate will be achieved. Such knowledge will be necessary in order to devise a robust technical and institutional system for sequestering CO_2.

THE ICM TOOL BOX

Industrial carbon management will be built atop the existing fossil fuel infrastructure. Building an effective ICM system will require adaptation and improvement of existing technologies as well as the development of new technologies to fill the gaps. We have a box of tools that have been proved by previous experience. We could assemble these tools today, with minimum modification, to build an ICM infrastructure for the production of electricity and hydrogen, but the cost of CO_2 mitigation would be relatively high, perhaps $100 to $250/tC. With the design of a few new components and careful optimization of components in integrated systems, it seems reasonable to suppose that the cost could fall substantially.

One can argue that the costs of many low-CO_2-emission technologies could be reduced with a bit more R&D, so how is ICM different? The answer lies in the close connection between ICM and the existing energy infrastructure and consequently in the scale at which the enabling technologies already exist. Consider four key examples:

1. *Coal gasification.* Integrated gasification combined cycle (IGCC) electric generation, a point of departure for many coal-based ICM designs, has not been adopted commercially despite decades of R&D; nevertheless, there is a large fleet of gasifiers now in operation with a worldwide syngas capacity equivalent to 50 GW thermal.

2. *Hydrogen production.* Steam methane reforming to produce hydrogen now consumes almost 2% of U.S. primary energy. Leading uses for the hydrogen are the production of ammonia and the reformulation of gasoline. Some current applications involve the long-range transport of hydrogen in pipelines.

3. *EOR.* In the United States about 7 MtC per year of CO_2 is used for enhanced oil recovery, most of it is supplied by pipeline from natural CO_2 formations. The longest pipeline runs 800 km. If the CO_2 were derived from fossil fuels, it would account for about 0.5% of U.S. primary energy.

4. *CO2 capture with MEA.* Monoethanolamine (MEA) solvents are used today for CO_2 capture from exhaust gases in more than 10 facilities worldwide and are also widely used for striping CO_2 from natural gas.

In addition to these enabling technologies, several integrated systems are important examples of nascent ICM. The first large project that sequesters CO_2 to avoid emissions is in Norway, where Statoil operates one of largest gas fields in Europe in which the produced gas contains about 10% CO_2, which must be reduced to 2.5% for sale to customers. The CO_2 is separated on an offshore platform and injected into a high-permeability aquifer under the seabed. The offshore capture and sequestration project was developed in response to Norway's offshore carbon tax of $170/tC. Statoil has sequestered about 0.3 MtC per year of CO_2 since 1996, and the transport of the sequestered CO_2 is now being monitored by an international research team (Herzog et al., 2000). A similar project planned in Indonesia at the Natuna field will inject 30 MtC per year, roughly 0.5% of present global emissions.

A project using the CO_2 from the Dakota Gasification plant to enhance oil recovery at Pan Canadian's Weyburn field is (arguably) the existing project that most resembles ICM. The Dakota plant has produced synthetic natural gas from coal since 1984. It is the largest facility of its kind and was a product of the Synfuels programs of the 1970s. Weyburn is a large oil field in Saskatchewan that is nearing the end of conventional production; with CO_2 EOR, the amount of recoverable oil will be increased by about 30%. A 325-km pipeline now transports 0.5 MtC of CO_2 per year from the gasification plant to Weyburn (Hattenback et al., 1999).

It is instructive to compare the scale of these technologies with current low-CO_2-emission alternatives. Nuclear and biomass are both used at very large scale, accounting for 9% and 3% of U.S. primary energy, respectively, but solar and wind are much smaller—0.08% and 0.04% respectively (Energy Information Administration, 1998). In comparison the Dakota Weyburn project alone sequesters 0.03% of U.S. CO_2 emissions, and all of the key enabling technologies listed above are in use at scales that far exceed the current scale of wind and solar.

ICM AND MITIGATION OF CO_2 EMISSIONS: TECHNICAL AND ECONOMIC BARRIERS

Industrial carbon management may be used to mitigate CO_2 emissions throughout most of the energy system; however, the heterogeneity of energy distribution and use means that the comparative advantage of ICM over other CO_2 mitigation technologies will vary widely. Most analysis of ICM has focused on electricity generation, but ICM can also be used to produce hydrogen, enabling deep reductions in CO_2 emissions via the substitution of hydrogen for natural gas or petroleum.

There are several good reasons to focus on electricity generation as an early application of ICM technologies:

- Electric power plants are among the largest point sources of CO_2.
- New electric generating technologies can be introduced without affecting the end user (other than by changing the cost of production).

• Most coal is used for electric generation (93% in the United States), and coal has the highest carbon-to-energy ratio of the fossil fuels.

Given existing technologies, the cost of electricity with ICM is estimated to be about 5-7 cents/kWh, about 2-3 cents/kWh more that the cost from current technologies (Herzog et al., 1997), and roughly comparable to the cost of electricity from wind, biomass, or nuclear power. Figure 8.3 illustrates the relationship between cost of electricity and intensity of CO_2 emissions for various technologies. For new coal plants, the cost of reducing CO_2 emission is about $100/tC. It is difficult, however, to estimate the real cost of reducing emissions from static cost comparisons because in real electric markets the introduction of ICM competes with fuel switching (coal to natural gas) and depends on the dynamics of plant dispatch. Figure 8.4 illustrates the cost of emissions reduction in a simple dynamic electric supply model.

Given open competition between electricity technologies under a carbon tax (or equivalent regulatory mechanism) and the assumption that carbon sequestration can meet environmental permitting requirements, there are significant structural reasons to expect sequestration to be adopted in preference to nonfossil alternatives even if the cost of electricity were similar. Unlike wind power and other distributed renewables, ICM plants would match the existing distribution system with respect to sizing

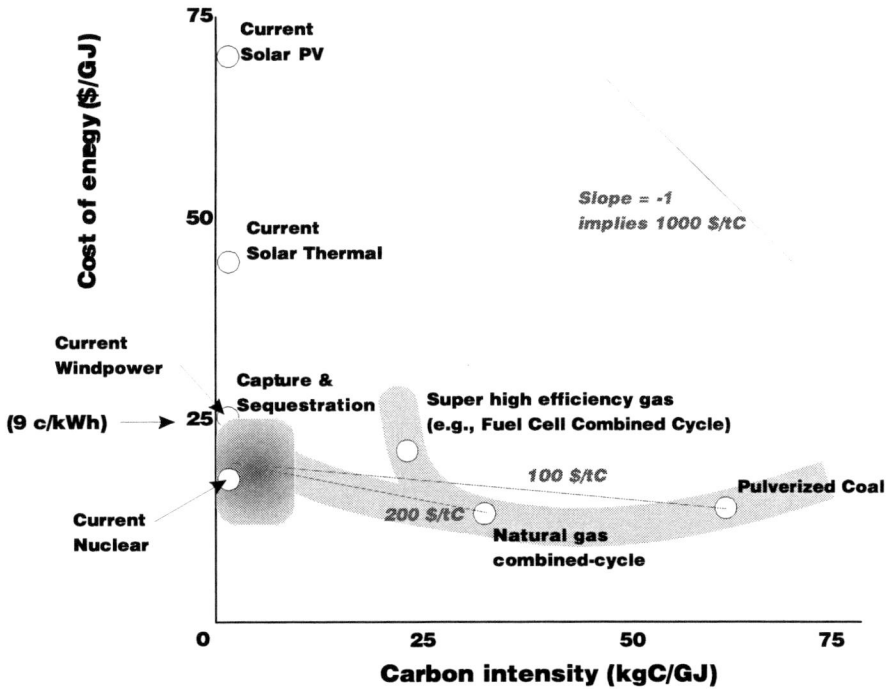

FIGURE 8.3 The cost of electricity versus carbon intensity. The x-axis shows CO_2 emissions (in kilograms of carbon per unit of electricity generation [in gigajoules]). The y-axis shows the approximate cost of electricity from new generating units including costs for capital, fuel, and operations. The likely cost of ICM technologies is on par with the estimated cost for large-scale wind or new nuclear. Currently, coal dominates fossil electricity supply, so replacement of coal with natural gas-fired electrical generation achieves substantial CO_2 mitigation at minimal cost, but this effect depends strongly on the price of natural gas as shown in Figure 8.4. Costs of intermittent renewables do not reflect the additional costs, such as storage, due to their intermittency.

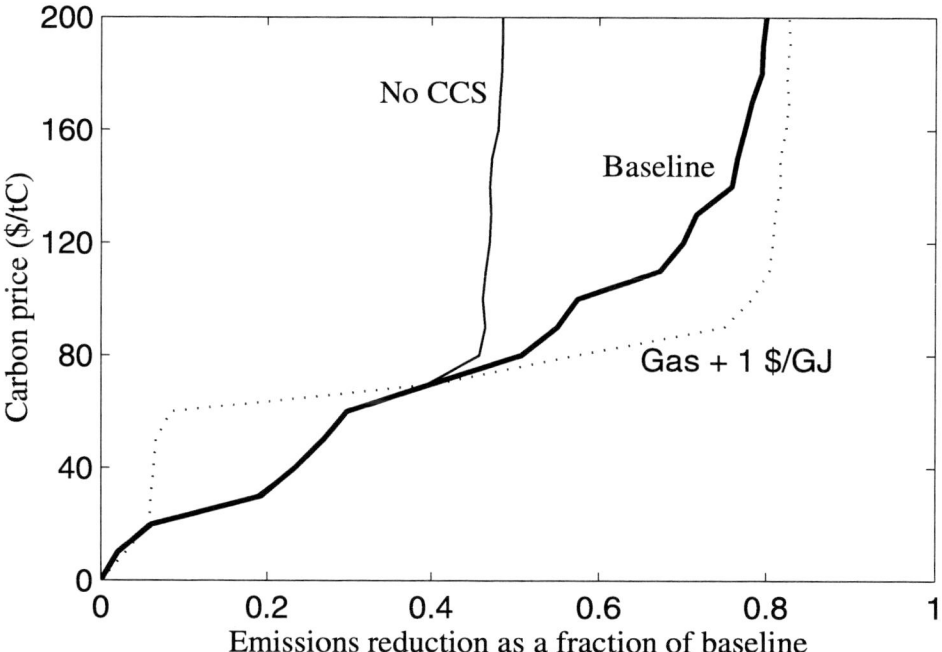

FIGURE 8.4 Carbon mitigation supply curves. These results are derived from the author's dynamic optimization model applied to a representative U.S. electricity market. Each point on the graph reflects the decrease in carbon emissions compared to a baseline scenario due to an imposed carbon price (set by a tax or equivalent regulatory mechanism). The "NO CCS" curve shows a model run without ICM technologies in which emissions decline by a maximum of about 45% due to conversion of all generation from coal to gas. The dotted curve shows the effect of increasing the price of gas by $1/GJ, raising the cost of moderate emission abatement by making coal-to-gas fuel switching less advantageous. NOTE: CCS = carbon capture and sequestration.

and ease of dispatch. Moreover, ICM plants with likely be constructed with existing suppliers, and established upstream fossil energy companies (the oil majors) could provide both fuel and CO_2 sequestration. While nuclear power could, in principle, play a central role in reducing CO_2 emissions, absent sweeping changes in the industry, its regulators, and public perception, it seems likely that utilities would find ICM less risky.

In sharp contrast to the introduction of new electric generating technologies, the introduction of hydrogen into dispersed stationary uses requires the development of a hydrogen distribution infrastructure—a serious challenge—while the introduction of hydrogen into transportation requires the development of effective hydrogen-fueled vehicles and a refueling infrastructure—likely an even greater challenge. Yet as a means to mitigate CO_2 emissions, the potential advantage of ICM over nonfossil energy sources (other than biomass) is due to the intrinsic advantages of thermochemical over electrochemical production of hydrogen. A crude comparison of energy costs serves to illustrate the point. At current prices, coal and natural gas, likely the most important fuels for ICM systems, have energy costs of roughly $1 and $3/GJ, respectively. Either feedstock can be used to generate electricity with ICM at a producer cost of $15 to $25/GJ (5-7 cents/kWh). As noted above, the cost of electricity produced from wind (absent all subsidies) arguably lies in the same range. In contrast, the price of H_2 produced from wind via electrolysis would be $20 to $30/GJ, while the price of H_2 produced from fossil fuels via ICM would be $4 to $7/GJ; a relative cost advantage of roughly 1:3 for ICM-hydrogen over wind-hydrogen despite the assumed equality of electricity costs. Similar disadvantages apply to the production of hydrogen from nuclear or solar power, though not from biomass. Moreover, large-scale production of hydrogen from fossil fuel and its long-range transport are already mature technologies in the petrochemical industry.

The relative ease of producing hydrogen via ICM implies that wherever hydrogen could replace oil or natural gas, the potential exists for comparatively inexpensive mitigation of CO_2 emissions. Realizing this potential, however, will not be easy. Substantial technical and economic barriers will hinder the diffusion of hydrogen fuel technologies across the energy system. Technical barriers range from the comparatively straightforward problems of constructing hydrogen-capable gas distribution systems to the serious engineering challenges that stand in the way of hydrogen-powered transportation systems. The economic barriers—including both economies of scale and network effects—are no less daunting. Consider the introduction of hydrogen-capable distribution systems; even if costs were low for both the distribution system and the end-user technology, the introduction of new distribution systems will likely be slow because distribution and end-user equipment must evolve together against the economy-of-scale advantages of existing systems.

Global Models

Although we have a limited understanding of the role of ICM technologies in bringing down the cost of CO_2 mitigation in the electric sector, very little is known about the influence of ICM on the overall cost of mitigating climate change. The uncertainty arises from the need to combine global economic models with models of technological change in time scales of the order of a century. Looking back 30 years at previous attempts to model the evolution of energy systems does not inspire confidence. Forecasting technological change would be difficult enough if one wanted to predict the evolution of a single technology, such as large-scale electric power generation. Predicting technological change over century time scales is still harder, however, because clusters of technologies evolve as tightly coupled systems and the evolution of the full system is highly path dependent.

Two extreme scenarios for the future of centralized electric generation serve to illustrate this path dependence. First suppose that electric power generation is rapidly decentralized, driven by the diffusion of small natural gas-fired combined heat-and-power generators—the technology that offers the most cost- effective near-term CO_2 mitigation. This would initially prevent diffusion of ICM because CO_2 cannot be effectively collected from distributed sources, but it would enable a later wave of decarbonization as ICM hydrogen (produced from cheap coal) competed against expensive natural gas. Alternatively, the economies of scale in large ICM electric generation might lower the *relative* cost of electricity—under a system-wide carbon tax—and cause acceleration of the fraction of primary energy converted to electricity at centralized facilities.

Despite the daunting challenges, several groups have used integrated assessment models (see Parson and Fisher Vanden, 1997, for a review of such models) to study the effect of ICM on the overall costs of stabilizing CO_2 concentrations. These models allow one to compute the reduction in CO_2 emissions resulting from imposing a price on emissions that approximates the effect of a carbon tax or similar regulatory mechanism. The models may be used to find a trajectory of carbon price over time that most efficiently stabilizes CO_2 concentrations at a given level. Conventional economic models suggest that peak marginal carbon prices of order $500 to $1,000/tC will be necessary to stabilize CO_2 concentrations at around 450 parts per million (ppm).[4] As we have seen above, simple technology cost

---

[4]Including other radiative forcings, 450 ppm CO_2 is approximately equivalent to a doubling of CO_2 over preanthropogenic levels.

estimates suggest that ICM could be used to mitigate a substantial fraction of total CO_2 emissions at much lower costs. The global model results in Figure 8.5 show decreases in mitigation cost of roughly a factor of two when ICM is included.

IMPLICATIONS

The oil crisis of the early 1970s intensified concerns that the world would soon run short of fossil fuels. Energy experts theorized that a global transition to nonfossil energy would be necessary within decades. Three decades later, although new discoveries and new recovery technologies have increased estimated fossil reserves and put to rest fears of their rapid exhaustion, concern about climate change has again led many experts to conclude that a rapid transition to nonfossil energy is required.

Part of the reason fears of oil scarcity proved exaggerated was that analysts failed to anticipate the potential for technical and managerial innovation to drive down the cost of petroleum exploration and extraction. We may have made a similar error in considering the link between fossil fuel use and climate. It has been assumed that the transfer of carbon from geologically isolated fossil reservoirs to the biosphere was a fundamental geophysical consequence of fossil energy use. Geological sequestration of CO_2 negates this assumption and raises the prospect that the long history of technical mitigation of the environmental impacts of fossil fuels can be extended to the climate problem.

By weakening the link between fossil energy and CO_2 emissions, carbon management makes it feasible to consider a fossil-based global economy through the next century, even in a greenhouse-constrained world. By reducing the severity of the threat that emission reduction poses to fossil industries and fossil-rich nations, carbon management may ease current political deadlocks. Stated bluntly, if carbon management is widely adopted and if existing fossil energy industries can extend their dominance into the new markets for carbon sequestration, then the increase in total energy costs will benefit industries that would otherwise lose by actions to abate emissions.

It is likely that carbon management will be a profoundly divisive issue for environmentalists. It may be opposed for at least two reasons. First, carbon management is only as good as the reservoirs in which the carbon is sequestered. If CO_2 leaks out much more quickly than we expect, then we leave our descendants with the double problem of uncontrollably rising CO_2 emissions and an economy still dependent on fossil energy. The history of toxic and nuclear waste disposal gives reason to be skeptical of expert claims about the longevity of underground disposal. Second, carbon management is a technical fix on a grand scale. It was first proposed as "geoengineering," a term now shared by proposals to engineer the global climate, for example, by cooling the planet by the injection of aerosols into the stratosphere to reflect solar radiation (Keith, 2001). In addition to a reasonable distaste for technical fixes, carbon management collides with the deeply rooted assumption among many environmentalists that fossil fuels are the "problem" and that renewable energy is the "solution." Yet, the rationale for support of carbon management is also strong. It may be that large-scale adoption of carbon management will allow the world to make aggressive CO_2 emissions cuts at a politically acceptable cost.[5]

ACKNOWLEDGMENTS

This research was supported by the Center for Integrated Study of the Human Dimensions of Global Change, a joint creation of the National Science Foundation (SBR-9521914) and Carnegie Mellon

---

[5] Several review articles cover much of the material presented here (Herzog et al., 1997, 2000, Socolow, 1997; Parson and Keith, 1998; Freund, 2000).

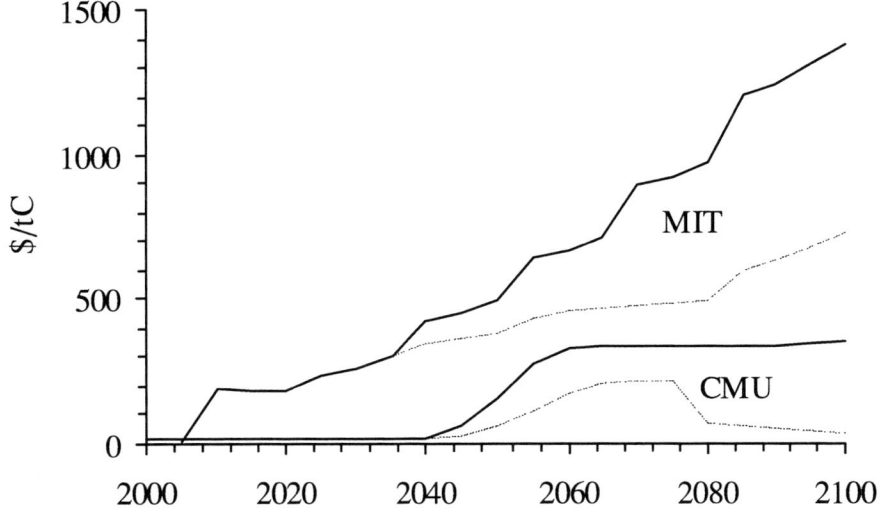

FIGURE 8.5 The effect of ICM on the global cost of stabilizing CO_2 concentrations. The y-axis shows the carbon price (set by a tax or equivalent regulatory mechanism) required to keep atmospheric CO_2 concentrations below about 550 ppm (twice preanthropogenic levels). Results from two very different models are shown together. In each case the basline simulation (solid line) does not include ICM, and the dotted line shows the carbon price if ICM technologies are included. The MIT model is a large general equilibrium economic model that reflects current technology and observed elasticities in demands for commodities (Biggs, 2000). The CMU model is a simpler economic model that includes parameterizations for technological change in response to price signals. The large disagreement between the estimated control costs reflects differing assumptions about technological change, and illustrates the great uncertainty inherent in such predictions.

University with additional support from the Department of Energy, Environmental Protection Agency, National Oceanic Atmospheric Association, Electric Power Research Institute, Exxon, and Applied Precision Inc.

REFERENCES

1. Biggs, S.D. 2000. Sequestering carbon from power plants: the jury is still out. In *Energy Laboratory*. Cambridge, Mass: Massachusetts Institute of Technology.
2. Energy Information Administration. 1998. *Annual Energy Review*. Washington, D.C.: U.S. Government Printing Office.
3. Freund, P. 2000. Progress in understanding the potential role of CO_2 storage. In *5th Conference on Greenhouse Gas Control Technology*, Cairns, Australia.
4. Hattenbach, R.P., M. Wilson, and K.R. Browncep. 1999. Capture of carbon dioxide from coal combustion and its utilization for enhanced oil recovery. Pp. 217-221 in *Greenhouse Gas Control Technologies: Proceedings of the 4th International Conference*, P. Reimer and B. Eliassen, Editors. Amsterdam: Pergamon.
5. Herzog, H., E. Drake, and E. Adams. 1997. *CO2 Capture, Reuse, and Storage Technologies for Mitigating Global Climate Change*. Department of Energy. Washington, D.C.: U.S. Government Printing Office.
6. Herzog, H., B. Eliasson, and O. Kaarstad. 2000. Capturing greenhouse gases. *Scientific American* 282(2):72-79.
7. Keith, D.W. 2001. Geoengineering. *Nature* 409:420.
8. Lackner, K. et al. 1995. Carbon dioxide disposal in carbonate minerals. *Energy* 20:1153-1170.
9. National Academy of Sciences. 1977. *Energy and Climate. Studies in Geophysics*, Washington, D.C.
10. Ogden, J.M., 1999. Prospects for building a hydrogen energy infrastructure. Pp. 227-279 in *Annual Review of Energy and Environment*, R. Socolow, Editor.

11. Parson, E., and K. Fisher Vanden. 1997. Integrated assessment models of global climate change. *Ann. Rev. Energy and Environ* 22:589-628.
12. Parson, E.A., and D.W. Keith. 1998. Fossil fuels without CO_2 emissions. *Science* 282(5391):1053-1054.
13. President's Science Advisory Committee. 1965. Restoring the Quality of Our Environment. Washington, D.C.: Executive Office of the President.
14. Rogner, H.H. 1997. An assessment of world hydrocarbon resources. *Annual Review of Energy and the Environment* 22:217-262.
15. Socolow, R.H. 1997. *Fuels Decarbonization and Carbon Sequestration: Report of a Workshop*. Princeton, N.J.: Princeton University.

DISCUSSION

Dave Cole, Oak Ridge National Laboratory: I wonder if you might elaborate on two points regarding the environmental aspect. First, elaborate a little bit on this issue of ocean sequestration and then on geologic storage of CO_2. Where do the environmentalists stand on these issues?

David Keith: With a few exceptions, CO_2 capture and sequestration is just are not on the radar screens of the environmental groups. But that is beginning to change. Several of the large environmental nongovernmental organizations NGOs know they need to grapple with CO_2 sequestration.

Concerning ocean sequestration, Norway is now separating out the roughly 10% of CO_2 that is in the gas on an off-shore platform from its largest gas field and shoving it down into a sandstone aquifer, whereas otherwise they would have been venting it. I have heard rumors that Greenpeace is going to try and get some other country to bring suit in the World Court to stop it because they believe it breaks the London Dumping Convention. I think, from my cursory reading, that Greenpeace is correct, that it does break the London Dumping Convention, because the Convention prohibits dumping industrial waste, not just in the ocean, but under the seafloor. If it's correct that Greenpeace would like to block Norwegian project, which in my view is an obvious environmental benefit, then that gives you a sense of how strongly people feel about the ocean.

I should add that not only has there been a huge Japanese research investment in oceanic sequestration, but there is a project that will run next year, funded by (I think) the Canadians, the Americans, and Japanese that will actually attempt to inject CO_2 at a rate of a kilogram per second. The project will run off Hawaii and will be used to model plume formation and so on. This experiment has got environmental groups to think about, and critique the idea of ocean sequestration.

Dave Cole: Just one quick comment to follow on that. The federal government is spending a fair bit of money examining the issue of ocean sequestration in terms of the geoenvironment.

David Keith: The overall budget for CO_2 mitigation is still small. As Bob Frosch said at a meeting we had at Harvard two years ago, we need to add a couple of zeros.

Dave Cole: Yes, I understand, but nonetheless, on these issues of policy and issues of environmental concern, I am concerned about the feedback between community interest or lack thereof and what really guides the research that we perform or the research that the federal government supports. There seems to be somewhat of a disconnect here.

David Keith: I couldn't agree more. I don't have any magic answer. There are more than a few disconnects both on the climate science side and on mitigation technology.

Dave Cole: The last part of my question concerns terrestrial sequestration. Are we going to encounter the same kinds of Greenpeace environmental concerns there—where people are actually within walking distance of where this stuff is being dumped?

David Keith: We just don't know because it is so early, but I think the answer is yes, but not as much for terrestrial as for oceanic CO_2 sequestration. At Carnegie Mellon, we had done a preliminary open-ended elicitation of knowledge and opinions about sequestration, with about 10 non-technical members of the community. While it is not representative, we found divergent views on geological sequestration where we found strong resistance to ocean sequestration.

Several folks I have talked to in the environmental community are willing to talk seriously about geological sequestration. But it is still in its early days. There are still no attempts at systematic risk assessment, and there is little sense of how, where, under what regulatory framework, and with what incentive sequestration would actually occur. All of these factors will likely influence people's reaction to geological sequestration.

Dan DuBois, National Renewable Energy Laboratory: I am a little more supportive of renewable energy than you are, but on the issue of how much CO_2 you could use, I think you are saying 0.1% or something like this. I think you misunderstand what people who worked with CO_2 would really like to do. They want to use CO_2 as an energy vector, much as you would use water splitting. When the CO_2 community talks about making a fuel from CO_2, we want to do much the same thing, only we want to make a liquid fuel. So if you have a nuclear source or a renewable energy source of electricity in the future, then you could provide not only hydrogen, which is a gaseous fuel, but also methanol a liquid—high-energy-density liquid fuels as well.

David Keith: I am quite skeptical that synthesis of fuels from CO_2 with energy provided by nuclear or carbon-free renewables can ever compete. But it is clear that we need systematic analysis that would come from some of you in the chemical sciences community. You can't arbitrarily start with nuclear or solar PV energy, which costs a lot, and only think about the subsequent analysis. You have to do a real analysis, which starts with fossil-free energy at a certain price or fossil energy at a certain price.

Dan DuBois: The question is, What price are you going to use and what assumptions are you going to make? This is a problem for 50 years in the future. You are saying that economic analyses 50 years in the future are very difficult. If you say that, on what basis are you going to make this economic analysis?

David Keith: All I am saying is that we should produce some robust analyses with transparent assumptions that address both sides of the problem. You clearly see the choice between either offsetting emissions, that is, using nonfossil energy to simply displace CO_2, or going to the CO_2 route to make fuels. A nice study like that could realistically be done with several assumptions, and if they are clearly laid out, would be a big help to DOE managers.

Panel Discussion

John Turner, National Renewable Energy Laboratory: Let me comment first on life-cycle assessment. The National Renewable Energy Lab (NREL) is doing life-cycle assessment for all renewable energy technologies, including photovoltaic (PV), wind, hydrogen—the whole thing, cradle to grave, energy balance, materials balance, all those things. The numbers are going to be available, and they are complete in a number of different issues. I think they have also done the biomass one.

Next, in terms of renewable energy for developing countries, India is one of the largest wind users in the world now and it is growing at a very high rate. One of the reasons for this is that even though renewable energy has a high initial capital cost, it is incremental. So, you can afford small increments at a time, and this helps developing countries implement renewable energy technologies. It is particularly true of wind.

It turns out that it is also true of PV. Even in India, people can afford small PV units, which give them a few hours of light in the evening. These units typically cost four or five hundred dollars and somehow they come up with the money. So, I think renewable energy technologies for developing nations are very useful because they can put them in at small quantities and afford them that way.

David Thomas, BP-Amoco: A couple of weeks ago I participated in another National Research Council (NRC) workshop on carbon dioxide sequestration that was coming at it from the biological point of view, which is terrestrial sequestration—soil carbon and standing trees and shrubs and so on. The real question to Dave Keith and perhaps the others that are involved in renewables is, Where do you see this fitting into the idea of sequestration? Terrestrial sequestration is driven primarily by the Department of Agriculture. It was talking about five- and ten-year contracts with the farmer to, "sequester carbon in his soil or on his land." What are your thoughts about this is the general question?

David Keith, Carnegie Mellon University: It seems increasingly clear that you can get a lot of bang for your buck for a while. What I mean by this is that it is clear that by reasonably inexpensive changes especially to agricultural practices and also to forestry, you can suck up a significant amount of CO_2. The new Intergovernmental Panel on Climate Change (IPCC) report, not yet released, will say something like a gigaton a year by such methods, which is a big deal clearly.

Of course, you don't suck it up for all time, and therein lie both a scientific and managerial problems. If farmers change their management practices so that they are sucking up a metric ton per hectare per year or something, this would be a really great number. However, carbon in soils is very labile. If you change the management practices, carbon can easily be oxidized back to CO_2.

So we need to invent financial instruments where there is some kind of perpetual lease and accounting procedure. A lot of the talk at the Conference of the Parties (COP) meeting in the last few weeks was about inventing such instruments. Everybody now agrees that we need to deal with the fact that you can't just pay somebody for their credit and then hope that it will stay there.

Effectively, I think, and many would agree, that we are going to need several colors of poker chips if we are going to trade carbon, where one color—the kind of gold standard—is just not emitting anything; another color is geological sequestration, which we think will stay there for 10,000 years or longer; and another color is biological sequestration, which is very easy to do but stays there for time scales of decades to a century.

David Thomas: One of the concerns that I had after participating in that meeting was the fact that there were three congressional aides present, all of who were beholden to fairly influential senators from agricultural states. It concerns me that the feedback to Congress through that group is going to be very strongly toward soil sequestration as the way to go. This could be a real issue for those that are concerned about the long term, as we are.

David Keith: Yes.

Andrew Kaldor, ExxonMobil: Let me ask a question concerning hydrogen storage. It seems to me that the central issue in changing over to different technology is a significant improvement in hydrogen storage. It appears that very high-pressure hydrogen is useful in commercial systems, whereas it is probably particularly difficult for individual drivers' fueling their vehicles.

So, the question I have is, Do you think the research ideas for hydrogen storage in solids are adequate or should we be putting more resources into this, perhaps approaching hydrogen storage in a significantly different way?

James Spearot, General Motors: In regard to how much we need to store, our goal is basically 10 weight percent (wt%) hydrogen as a fraction of the total fuel system weight. We believe that at this level we could operate a fuel cell sport utility vehicle (SUV). We could give the customer what he wants; he could drive a large vehicle, pull a trailer, whatever, and would still have a reasonable range with 10% hydrogen storage fuel cell vehicle.

In terms of the options that are out there at the moment, I mentioned gaseous storage developments. Some of the tank suppliers using advanced lightweight carbon fiber tanks are getting to—or at least contemplating—10,000 pounds per square inch. This gets us into the 7% range.

Liquid cryogenic hydrogen storage would get us to our 10% range if we wanted to develop an infrastructure to pump cryogenic hydrogen around. Obviously, producing cryogenic hydrogen requires a lot of energy itself. So you really have to talk about a renewable energy source not only to create the hydrogen, but also to compress it and to liquefy it.

Liquid cryogenic hydrogen is a leading option in Germany at the present time. There are demonstration programs going on at present, and they believe that this is a true end game for their particular economy at some point in the future. We haven't reached the point where we believe the same thing.

We would like to see the development of high-weight-percent hydrogen storage. This includes solid absorption tanks. The other technologies that are out there right now include chemical hydrides that are complexed with water, which give a little extra kick in terms of hydrogen availability. Then, of course, there are the carbon materials, primarily carbon nanotubes, although we believe some of the things that have been reported in the literature may not be nanotubes.

Right now, some tremendous claims are being made. We are struggling to reproduce those claims, and I think this solid storage is a longer-range option. We will be well into the second decade of this century before we are going to have solid absorption tanks out there. They would represent the safest system that we could develop. The question is, Could either compressed or cryogenic be made safe? We are looking at this very carefully.

John Turner: Just to comment on hydrogen, we have to differentiate between two areas. You are talking about vehicle storage. Stationary storage of hydrogen is really not an issue. You can use the standard compressed tanks, or you can even pump it underground and do other sorts of things. So stationary hydrogen storage for long-term energy storage is not an issue. Storage of hydrogen on mobile vehicles is certainly an issue.

The numbers I saw recently out of Thiacol using its tank was more like 11 wt% at 5,000 psi. If you added all the system to this, it brings the percentage down, but that is the latest number I saw, about six months ago, when Thiacol announced its new tank.

In terms of carbon additives, I think the leaders are NREL. Mike Heben, in research funded by Honda, is showing 7% hydrogen storage with nanotubes, but he still has some issues with regard to manufacturability.

I think nanotubes will work. We have seen enough experimental data. The problem is how do you take something now that costs a thousand bucks a gram and turn it into a commodity chemical at $2 a kilogram, especially something as engineered as a carbon nanotube. This is a big issue right now.

David Keith: Another reason I am not that worried about hydrogen storage involves where in the transport infrastructure you would first introduce hydrogen. The conventional view, of course, is in personal automobiles. Our view at Carnegie Mellon is different. We suggest a focus on heavy transport systems, such as ships, heavy trucks, and rail.

There are several reasons. First, the storage problem is much reduced for heavy transport because there isn't such a volumetric constraint. Second, many of these modes are now very dirty with respect to conventional pollutants because we haven't regulated them hard. You want to pick up the benefit of mitigating both CO_2 and conventional pollutants if you are going to justify an expensive, risky technology. So, for example, ships account for 2% of global CO_2 emissions and 14% of global NO_x emissions because they have these incredible NO_x-producing engines. It is the same for sulfur. So, if you want to have a double environmental win, there is a reason to start on heavy transport modes than rather than on personal autos. Third, a lot of these technologies are in operation for most of the day, which means more effective amortization of capital cost. You can afford to spend more capital on an expensive technology that will lower your operating cost (under a carbon tax) if the technology utilization is high. Cars aren't used that many hours a day, so they are very sensitive to capital cost. If you are being rational about putting hydrogen into the transport system, I don't think you start with cars.

James Spearot: I basically agree with this. You will see some in cars, but they will be early prototype demonstrations and limited product applications to get the knowledge base developed, but clearly center-city buses make an ideal application for hydrogen fuel cells.

Richard Foust, Northern Arizona University: I have a question relating to policy about renewable energies. From what I remember, the price of solar photovoltaics in the late 1970s and early 1980s was not a lot different in real dollars from what it is right now. If we were to increase the amount of money directed to research, would it significantly advance the time that the prices of renewable energies would fall?

John Turner: For photovoltaics, it depends on the technology. If you wanted to talk about thin-film technologies, which have perhaps the highest possibility of really low costs in terms of their systems, then, yes, more funding for research would definitely be in order.

For single-crystal silicon, the problem is in manufacturing. Scaling-up manufacturing would decrease costs. I don't remember what the costs were in the 1970s and 1980s but right now if you are a large buyer, the parking lot I showed in Sacramento costs about $4 a watt installed.

Manufacturers are making panels. The lowest cost is about $1.50 per watt. The average was $2.73 last year. They are projected to come down as manufacturing volume goes up. However, thin-film technologies, which have the opportunity to lower the costs to 6 cents/kWh almost as soon as they go into full production, could certainly use a lot of research dollars. We don't understand not only how to make the technology work, but even why it works the way it does.

Richard Foust: So it is basic research money that needs to be spent, not engineering research?

John Turner: For thin-film photovoltaics, it is basic research money. For single-crystal silicon, it is engineering research and reducing manufacturing costs.

Richard Foust: Has the price dropped relative to funding for research?

John Turner: There is a program at the Department of Energy called PV Materials that has been a significant player in lowering the manufacturing costs of photovoltaic panels.

David Keith: It is clear that spending money can buy down the price of a lot of energy technologies, and for many, if you do historical studies and you plot the log of the total amount we built so far versus the cost, you get a nice line. It is a power law. This is true for gas turbines, and it is beautifully true for solar photovoltaics, but it is important to compare apples with apples—to compare technologies with the same assumptions. Analysts who wish to advocate a given technology often apply a learning curve model to it, but not to the competition. We are still far from a robust understanding of technological innovation and diffusion in the real world.

Klaus Lackner, Los Alamos National Laboratory: I would like to point out that in all of these discussions of what energy form is better or worse or which one we should use, our ultimate goal has been to take into consideration that there will be 10 billion people who would like to have a standard of living as high as we have today, and if we were to get there over the next hundred years, it would be a factor of 10 increase in energy consumption.

I would argue that we really need cheap, clean energy and we need a large amount of it. I think every little bit we can get to mustering this is worthwhile. I don't view alternate energy forms as competing approaches, but as multiple alternatives to get to the goal. I would argue that we cannot exclude any of them.

Fossil energy, on the one side, is 86% of the total right now. If we pull the rug out from under that, we would have a catastrophe. On the other side, I think renewables could grow a lot and make a big difference, particularly in the distributed energy market. The market is so large that if we can all get the prices down, we do a service to the world because people do need energy. I believe there are 2 billion people without electricity right now.

Robert Wilson, SRI: I would like to go back to the question of the utilization of biomass and its gasification of combined with sequestration. In all the discussions we have had today, we focused on land utilization issues. The question hasn't come up about utilizing ocean resources for farming biomass for the same purpose.

The second question is, David, you mentioned that gasification of this biomass was easy and I would like to know the basis for that assertion.

David Keith: "Easy" was the wrong word. I meant to say easier than coal. Even this may be overstated. While biomass has less problems with sulfur, less ash, and gasifies at lower temperatures, it has other challenges that compensate for many of these advantages.

The ocean answer is that the ocean productivity is not so big, and it is actually hard to extract substantial energies from the oceanic biota.

Robert Wilson: We are not talking about the natural productivity of the ocean. We are talking about farming—ocean farming.

David Keith: Farming, meaning fish farms?

Robert Wilson: No, well—

David Keith: Hard to do.

Robert Wilson: The growth rates of some of these ocean materials are far greater than the growth rate of some of the grasses that people are talking about today. So, this is why it is of interest to some people.

Alan Wolsky, Argonne National Laboratory: This is a comment prompted by some of the things that have been said, but the comment is really directed to the organizers of the meeting and the National Academies. Our field does not have a cumulative literature that would help in these discussions. For example, 20 years ago, people tried to grow kelp and they were successful because kelp grows like crazy, but 19 years ago, they found out that when a storm came they couldn't protect the kelp and they lost their crop. This is a factoid I know about, but the absence of a general knowledge of it illustrates the lack of continuity in our literature.

The difference between the science and this quasi-policy discussion is that in science, there are usually indices and a way to benefit from the work already done. In policy I know many cubic feet of reports—and many of these cubic feet discuss topics we are touching on here—but we have failed to benefit from past work, just as our discussions may not be accessible to the future. So just to do one more study without thinking about how the present interest in climate change can be made accessible to those who come after us misses an opportunity to really make a contribution.

Tom Baker, Los Alamos National Laboratory: The thing that has been disappointing for chemists in the issue of carbon sequestration has been the lack of interest in the United States in expanding funding for the fundamental chemistry of carbon dioxide. In Europe, there is a lot of focus on CO_2 reuse and recycling. Here, instead, the attitude has been that it is never going to be large enough to make a difference in the sequestration area, so let's spend our money elsewhere where we are going to solve the larger problem. In fact, there is a lot of chemistry that needs to be worked out with CO_2. We still really don't know much about the very basic fundamentals, and in spite of the thermodynamics, there is still a lot of chemistry that could be done with higher-energy coreactants. So how do you think we could go about getting the United States to pay a little more attention to the fundamental chemistry of CO_2?

Carol Creutz, Brookhaven National Laboratory: I agree with your assessment and do not have any new ideas on how to focus attention on fundamental CO_2 chemistry. I would also comment that within even the Office of Science at the Department of Energy, there seems to have been a much more dominant role of the Office of Biological and Environmental Research (OBER) in attacking the carbon management problem and identifying sequestration as the solution. I don't know if this reflects a whole community or the local response within the United States. Yet the problem has been taken on as a sequestration problem, which—to me as a chemist—doesn't make sense. It seems to me that there would be a lot of things we could be doing and benefiting from.

9

Managing Carbon Losses for Selective Oxidation Catalysis

Leo E. Manzer
DuPont Central Research and Development

Catalytic oxidations are among the least selective of all catalytic reactions and the source of much of the carbon dioxide from chemical processes. These processes are often carried out at high temperatures, resulting in selectivity for the desired oxygenated products of less than 90%. The major by-product is usually carbon dioxide. Table 9.1 shows typical selectivity for the major product in these large-scale commercial operations. Since most of these processes have production capacities of several hundred million pounds per year, the amount of CO_2 generated is quite significant. For example, for a malefic anhydride plant operating at a capacity of 200 million pounds per year, with 60% selectivity, more than 500 million pounds per year of carbon dioxide is produced per year. Therefore, there is a large incentive to improve yield of the desired hydrocarbon product in these processes. This can be done by improving the catalyst in these existing processes or by completely changing the chemistry and engineering of existing processes.

ANAEROBIC VERSUS AEROBIC OXIDATIONS

It has been known for many years that in certain oxidation reactions the most selective catalysis occurs when the oxygen in the final product is derived directly from the lattice of the oxide catalyst. However, most large-scale commercial processes are carried out in the nonflammable region, which usually requires a feed of less than 3% organic and the balance air. This large excess of gas-phase oxygen results in the low selectivities shown in Table 9.1. The Mars-van Krevelen mechanism (Figure 9.1) suggests that a gain in selectivity is possible by keeping gas-phase oxygen from the process. The lattice oxygen of the catalyst is used in a stoichiometric reaction with a hydrocarbon to yield the oxygenated product. The reduced oxide catalyst is then transported to a separate zone and reoxidized by air. The process is referred to as *anaerobic oxidation.*

For an anaerobic oxidation process to be successful and economically viable, several requirements must be met. First, there must be a selectivity improvement for the desired product by operating in this mode. Next, the catalyst must have a high oxygen-carrying capacity per unit weight, to minimize the amount of catalyst circulated through the reactor. This is very important because the catalyst is often

TABLE 9.1 Selectivity to Major Product of a Few Commercial Catalytic Oxidation Reactions

Oxidation Process	Major Product	Selectivity (%)
Butane oxidation	Malefic anhydride	60
Propylene oxidation	Acrolein or acrylic Acid	75
Propylene ammoxidation	Acrylonitrile	80
Ethylene oxidation	Ethylene oxide	88

more expensive than the oxygenated product. Reoxidation of the reduced catalyst in the regenerator should occur at a temperature similar to that of the oxidation step to minimize the need to heat or cool the catalyst solids (energy minimization). Finally, since the reaction is essentially stoichiometric between catalyst and organic, a large amount of solid must be circulated around the reaction system. The catalyst must therefore be very resistant to attrition and must maintain structural integrity through many redox cycles. An excellent, early example of a two-step, anaerobic oxidation is the Lummus process for ammoxidation of o-xylene to o-phthalonitrile (dinitriles, DNs).[2] A simplified schematic is shown in Figure 9.2. A higher selectivity is claimed for the two-step process relative to the single-stage, aerobic oxidation.

Another large development effort was carried out by ARCO Chemical during the 1970s to convert methane to ethylene.[3] The reaction occurs at a very high temperature of 850-900°C. Patent and literature references, which illustrate the use of a $Li_{0.5}B_{0.5}MnMg_{2.8}O_x/SiO_2$ catalyst, show that at a conversion of 22%, the selectivity to two-carbon compounds was about 60% under both aerobic and anaerobic conditions. However, the yield of CO_2 was reduced from 34% to 22% when the reaction was conducted in a cyclic mode, by carrying out the oxidation under anaerobic conditions. A major development effort was terminated when the price of oil decreased.

Scientists and engineers at Monsanto studied the oxidative dimerization of toluene to stilbene,[4] as part of a new styrene process (Figure 9.3). In the first step, using a $K_{0.43}BiO_x$ catalyst at 575°C, the anaerobic process showed higher conversion (46 vs. 38%) and higher selectivity to stilbene (81.3 vs. 72.7%).

Emig has recently studied the oxidative dimerization of isobutylene to 2,5-dimethylhexadiene (DMH).[5] Under aerobic conditions, a conversion of 24% was obtained with a selectivity of 38% to DMH giving a single-pass yield of 9.1%. Under anaerobic oxidation, the conversion dropped to 11%, the yield remained constant at 9.9%, and the selectivity to DMH increased from 38 to 90%. This is a remarkable example of CO_2 reduction using a two-step process.

A recent patent has described the oxidation of propylene to acrolein and acrylic acid using a multicomponent metal oxide catalyst in a circulating solids reactor (CSR).[6] Under anaerobic conditions

$$M_xO_y + C_4H_{10} \longrightarrow M_xO_{y-f} + C_4H_2O_3$$

Catalyst Regeneration Step:

$$M_xO_{y-f} + O_2 \longrightarrow M_xO_y$$

FIGURE 9.1 Mars-van Krevelen oxidation of butane to maleic anhydride.

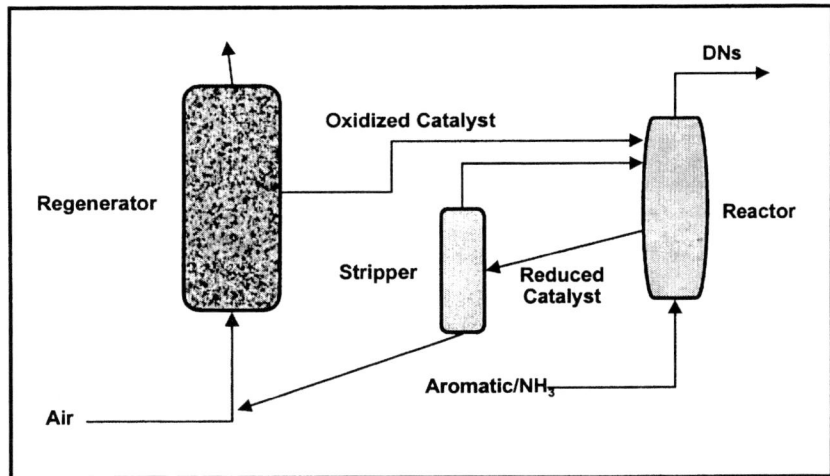

FIGURE 9.2 Simplified schematic for the two-stage oxidation of *o*-xylene.

at 350°C, propylene conversion was 16% and selectivity was 95.5%. When air was introduced into the CSR, the conversion increased to 21% and selectivity dropped to 82%, once again showing the substantial advantage of keeping gas-phase oxygen out of the catalytic oxidation zone.

DuPont recently commercialized a new process for the oxidation of butane to maleic anhydride using a CSR.[7] The maleic anhydride is scrubbed from the reaction zone as maleic acid and then hydrogenated to tetrahydrofuran. The advantages are well documented in the references. A key to this process was the development of an attrition-resistant catalyst obtained by spray-drying a solution of micronized vanadium-phosphorus-oxygen (VPO) catalysts in polysilicic acid.[8] In the spray dryer, a porous shell of very hard silica is formed to protect the soft VPO catalyst.

These few examples show an advantage of anaerobic oxidations for selected reactions, to minimize CO_2 formation. A few other opportunities for further study should include the oxidation of *o*-xylene to phthalic anhydride, oxidative dehydrogenation of ethylbenzene to styrene, oxidation of isobutylene to methacrolein and methacrylic acid, and oxidative dehydrogenation of paraffins to olefins.

PARAFFIN OXIDATIONS

Currently, there are no commercial processes involving the direct gas-phase oxidation of paraffins to an oxygenated product or olefin. The conventional approach involves endothermic dehydrogenation

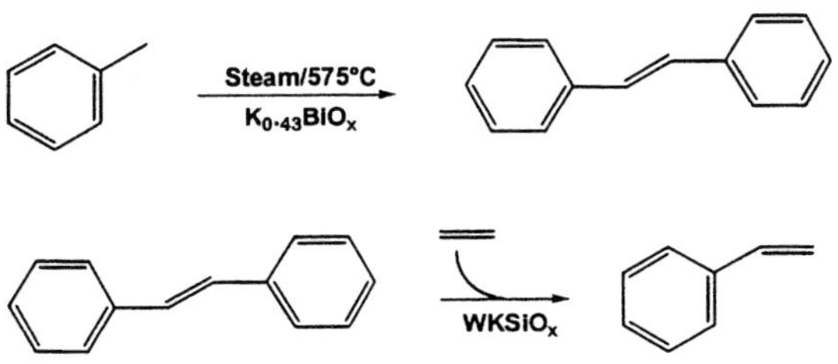

FIGURE 9.3 Monsanto anaerobic oxidative coupling of toluene.

of paraffin to the desired olefin followed by oxidation to the desired product. This process generates CO_2 in the endothermic dehydrogenation step because heat must be provided to the reaction. By comparison, if direct oxidation of the paraffin to the desired oxygenates could be achieved at high selectivity, there would be a net reduction in CO_2 and the process, in fact, would export energy (Figure 9.4). A number of companies have been very active in this area for many years.[9]

The results from selected patents for the ammoxidation of propane to acrylonitrile are shown in Table 9.2. Significant advances have been made over the past 10 years. High conversions and selectivities approaching 64% have been obtained. BP has announced that a pilot plant is operational to collect basic data for commercial design. The stake is high for this development because of the lower cost of propane compared to propylene and the reduction in CO_2 emissions.

Another well-studied process, in which significant progress has been made, involves the oxidation of propane to acrylic acid.[10] (See Table 9.3.) These results are quite impressive, with selectivity reported in excess of 80%. By contrast, direct catalytic oxidation of isobutane to methacrylic acid has been less developed.[11] Sumitomo has reported that 42% methacrylic acid can be obtained at 25% conversion.

NEW PROCESS CHEMISTRY OR CONDITIONS

Up to this point, the focus has been on improving the yield of the catalytic reaction to reduce CO_2 emission. However it is important to consider entirely new process chemistry that might reduce the number of steps, lower the temperature, and as a result, also lower CO_2 production. An excellent illustration of this point involves the production of methyl methacrylate (MMH). Current commercial catalytic routes use C_4 feedstocks and involve two high-temperature gas-phase catalytic steps followed by esterification. The first two steps occur above 350°C with an overall yield of about 75%. The main by-product is carbon dioxide. A new process to methyl methacrylate is under development by Asahi Chemical.[12] This process combines the second and third steps into a single oxidative esterification step

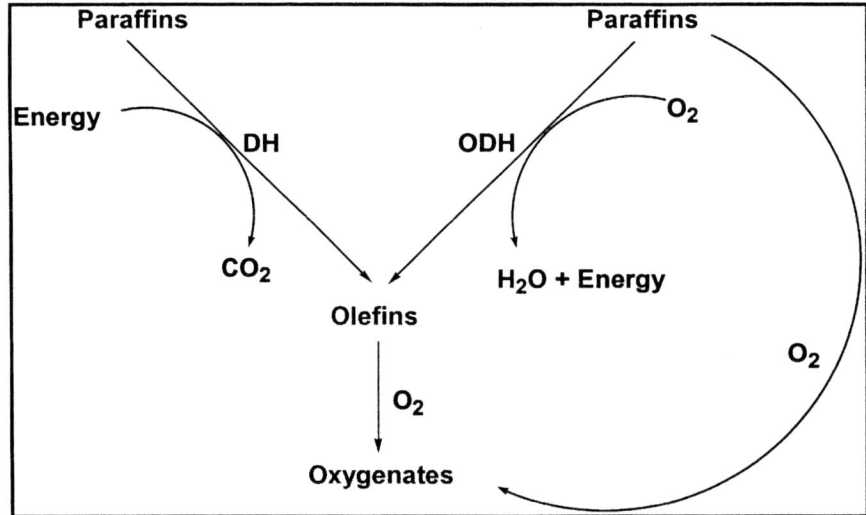

FIGURE 9.4 Incentive for direct oxidation of paraffins.

TABLE 9.2 Selected Results for Ammoxidation of Propane to Acrylonitrile

Company	Catalyst	Conversion (%)	Selectivity (%)
BP	$VSb_{1.4}Sn_{0.2}Ti_{0.1}O_x$	30.5	58.3
Asahi	$NbSb_aCr_bX_yO_n$	29.1	30.7
Mitsui Toatsu	$V_1Li_{0.1}P_{1.1}O_x$	54.8	58.8
Mitsubishi	$MoV_xTe_{0.2}Nb_{0.1}O_{4.25}$	79.4	63.5

(Figure 9.5). Using a Pd/Pb/Mg-Al$_2$O$_3$ catalyst, Asahi reports better than 98% conversion of the methacrolein to methacrylic acid with a selectivity greater than 95%. The reaction occurs at a mild 80°C in a slurry-phase reactor. The overall yield is significantly higher than in the conventional process, less CO$_2$ is generated, and capital investment is lower. A plant with a capacity of 135 million pounds per year is currently under construction. Mitsubishi Rayon[13] has also been active in this area with a Pd$_5$Bi$_2$Fe/CaCO$_3$ catalyst giving better than 97% selectivity for MMA at 76% conversion.

ALTERNATIVE OXIDANTS

Oxygen or air will likely be the preferred source of oxygen from an economic standpoint for many years. However, a growing number of developmental applications with hydrogen peroxide (H$_2$O$_2$), nitrous oxide (N$_2$O), and alkylhydroperoxides as the oxygen source are appearing in the literature. A relative comparison of costs of various oxygen sources is shown in Table 9.4. Clearly, from a cost standpoint, it will be difficult to justify new commodity chemical processes using on-purpose production of N$_2$O and H$_2$O$_2$ as oxygen source. However for fine chemical applications such as pharmaceuticals and agrochemicals, the cost may well be justified. Hydrogen and oxygen mixtures may be economically justified, although safety issues will likely require a greater investment.

Panov has extensively studied the use of N$_2$O as a selective oxidant for aromatics.[14] Using zeolites that contain only small amounts of iron, he has shown that benzene can be oxidized to phenol with selectivity of over 99% at around 300°C. Emig has studied the mechanism of this interesting reaction.[15] He proposes that at temperatures of less than 300°C in the absence of benzene, all the N$_2$O reacts with the surface to give an α-oxygen site that is very stable (Figure 9.6.) Cooling the solid below room temperature and introducing benzene give phenol in high selectivity. The lifetime of the site is 0.5 seconds at 500°C and 1.75 seconds at 420°C.

TABLE 9.3 Selected Results for the Oxidation of Propane to Acrylic Acid

Company	Catalyst	Conversion (%)	Selectivity for Acrylic Acid (%)
Wang (Fudan U)	$V1Zr_{0.5}P_{1.5}O_x$	18.3	81.0
	$Ce_{0.01}VPO$	27.2	68.3
Toa Gosei	V-Sb-Mo-NbO_x	30.9	29.4
Mitsubishi	$V_{0.3}Te_{0.23}Nb_{0.12}Bi_{0.017}MoO_x$	56.2	42.6
Rohm and Haas	$V/Te/Nb/MoO_x$	71.0	59.0

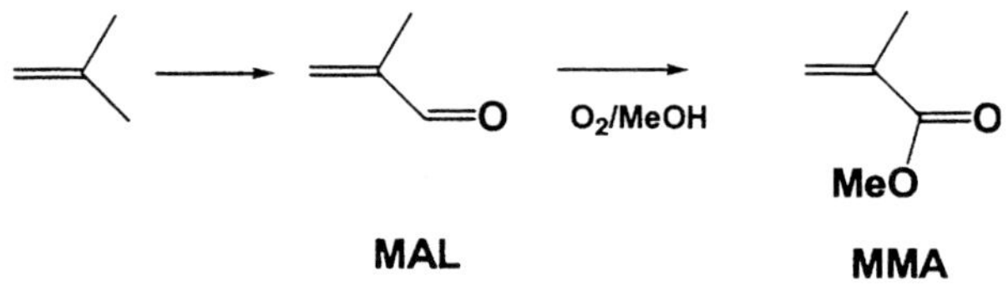

FIGURE 9.5 New route to methyl methacrylate. NOTE: MAL = methacrolein.

Through a very extensive collaboration with Solutia, this technology has been integrated into nylon intermediates manufacturing. Conventional technology for the production of adipic acid from cyclohexane provides cyclohexanone (K) and cylcohexanol (A) as intermediates. Nitric acid is used to oxidize K/A to adipic acid and in that step produces significant amounts of N_2O. Historically, the N_2O-containing gas stream has been vented to the atmosphere, but due to ozone depletion issues, most producers now abate the N_2O. Solutia decided to separate the N_2O after the nitric acid oxidation of K/A and to react it with benzene to produce phenol, which can be hydrogenated to K and oxidized to adipic acid. This provides an opportunity to use the N_2O for expansion purposes and to provide a higher incremental yield of adipic acid. The process is reported to be in pilot plant production.[16] Application of this technology for uses other than retrofit options will be highly dependent on the cost of N_2O. A recent patent to Solutia[17] involves the use of $Bi/Mn/AlO_x$ as a catalyst for the oxidation of ammonia to nitrous oxide. N_2O selectivity is reported to be about 92% at 99.2% conversion of NH_3. The cost of N_2O is projected to be about 25% that of H_2O_2, so if this new process is commercialized, it might be a serious alternative to O_2 oxidation.

Titanosilicalites[18] are now well known to oxidize a wide variety of organics with hydrogen peroxide or alkylhydroperoxides as the oxygen source (Figure 9.7.) Selectivity is generally very high, although utilization of the peroxide is often low. Because of the high cost of hydrogen peroxide (Table 9.4), commercial use will likely be restricted to fine chemical applications. However the use of hydrogen-oxygen mixtures is currently showing great promise.

For many years, DuPont studied the direct catalytic combination of hydrogen and oxygen to hydrogen peroxide.[19] The use of platinum and palladium bimetallic catalysts on silica was studied extensively. Selectivity for hydrogen peroxide was highly dependent on the weight ratio of platinum to total metal loading on silica. Optimum ratios were 0.02-0.2, which yielded selectivity for H_2O_2 of about 70%. Very high pressures were reported to be used and concentrations of hydrogen peroxide exceeded 20%. A key to achieving the high selectivity was the addition of a promoter such as Cl^- or Br^-.

Hoelderich has been studying the epoxidation of propylene to propylene oxide using titano- or vanadiosilicalite (VS) catalysts.[20] Using a titanosilicalite (TS) support with Pt/Pd and an NaBr promoter (Figure 9.8), with a mixture of hydrogen and oxygen, he was able to achieve selectivity for propylene

TABLE 9.4 Cost of Various Oxygen Sources

Oxygen Source	Cost ($/lb-mole)	Cost ($/lb)
O_2	0.64	0.02
N_2O	7.5	0.17
H_2O_2	17	0.50
$H_2 + O_2$	0.65	0.02

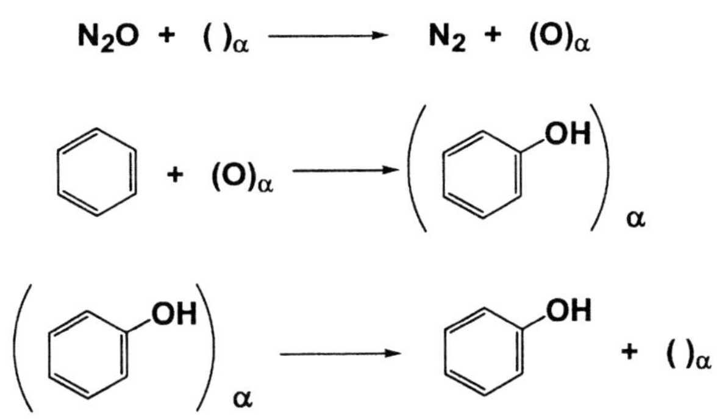

$$N_2O + (\ )_\alpha \longrightarrow N_2 + (O)_\alpha$$

$$\text{C}_6\text{H}_6 + (O)_\alpha \longrightarrow \left(\text{C}_6\text{H}_5\text{OH}\right)_\alpha$$

$$\left(\text{C}_6\text{H}_5\text{OH}\right)_\alpha \longrightarrow \text{C}_6\text{H}_5\text{OH} + (\ )_\alpha$$

FIGURE 9.6 Emig's proposed mechanism for the selective oxidation of aromatics by N_2O.

oxide of 87.3% at 19.4% conversion. In the absence of the NaBr promoter, selectivity dropped to 34.1%. The result is an exciting illustration of the potential for in situ production of a peroxo species as a selective oxidant.

Recently, carbon dioxide has been reported to act as a mild oxidant with chromium-based catalysts at very high temperatures. Longya et al. have studied the oxidative dehydrogenation of ethane using modified chromium catalysts that are support on silicate-2.[21] At 1073°K, good selectivity for ethylene is seen (Table 9.5.) An evaluation of the data suggested that the overall chemistry is very complex (Figure 9.9). Two successive coupling reactions occur. The first involves the dehydrogenation of ethane to ethylene and hydrogen. The second reaction involves the reverse water-gas shift reaction, to form CO and water, thus allowing for continuous removal of hydrogen during the dehydrogenation step. The process is reported to be commercial for the conversion of ethane to ethylene in an FCC (fluid catalytic cracking) tail-gas stream. The ethylene produced is acceptable for the formation of ethylbenzene.[22]

Recently Wang et al. have studied the effect of the support on the oxidative dehydrogenation of ethane with ethylene using carbon dioxide.[23] The data are shown in Table 9.6. All data are at 600°C. Silica appears to be the optimum support in this study as well. Very high selectivity for ethylene is seen.

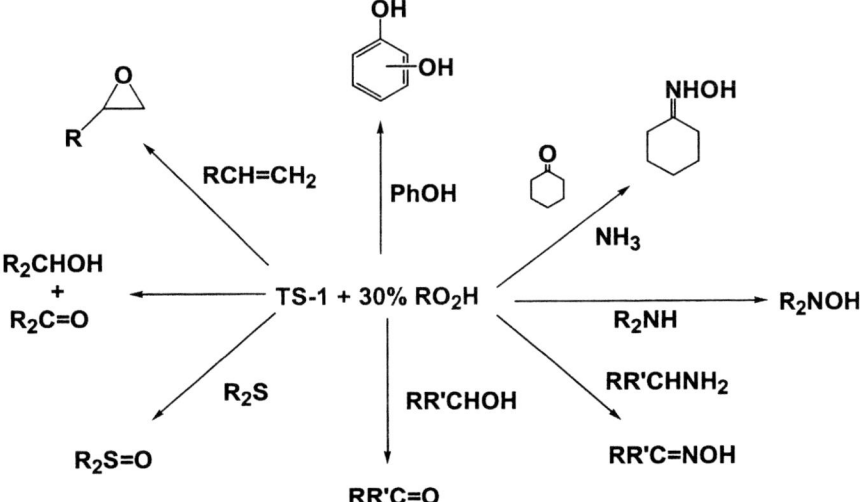

FIGURE 9.7 Range of products made by using titanosilicalites.

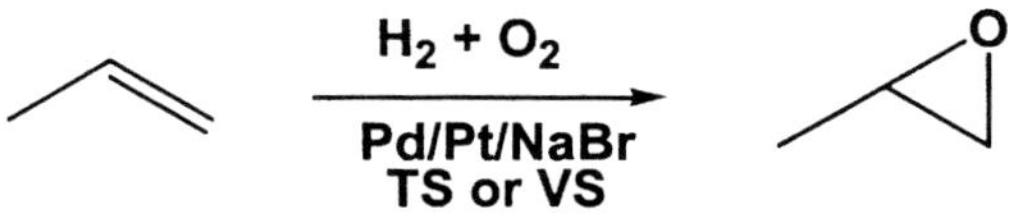

$$\text{CH}_2=\text{CHCH}_3 \xrightarrow[\text{TS or VS}]{\text{H}_2 + \text{O}_2, \text{ Pd/Pt/NaBr}} \text{propylene oxide}$$

FIGURE 9.8 Preparation of propylene oxide from hydrogen-oxygen mixtures.

CONCLUSIONS

A significant reduction in the amount of CO_2 that is emitted to the atmosphere is a desirable global goal. This workshop has suggested a number of opportunities to convert CO_2 into useful products. However, for chemistry, the ultimate goal is to eliminate the production of carbon dioxide completely in new processes. Clearly, it will be uneconomical to replace existing plant investment, so end-of-pipe treatment is important. The purpose of this chapter is to show that indeed there are a number of emerging new catalysts and catalytic processes that have higher selectivity for products and thus result in reduced CO_2 production. While most of these developments are only in the discovery or early development phase, there is sufficient progress to indicate that commercialization is a real possibility in the future. The following are the key points of this chapter:

1. Anaerobic oxidation of hydrocarbons can offer significant reduction of CO_2 in several cases. However, several criteria must be met for economical viability.

2. Alternative oxidants such as H_2O_2, RO_2H, and N_2O can provide higher selectivity in many reactions, but economics are currently attractive in only a few isolated cases.

TABLE 9.5 Oxidative Dehydrogenation of Ethane to Ethylene on Chromium Catalysts

Catalyst	Conversion (%) CO_2	Conversion (%) C_2H_6	Selectivity (wt%) CH_4	Selectivity (wt%) C_2H_4	Selectivity (wt%) H_2/CO
Cr/Si-2	18.6	58.9	19.6	80.4	1.4
Cr-Mn/Si-2	22.2	62.4	18.4	81.6	1.4
Cr-Mn-Ni/Si-2	24.2	67.9	18.7	81.3	1.6
Cr-Mn-Ni-La/Si-2	20.5	64.2	13.8	86.2	1.4

$$16 C_2H_6 + 9 CO_2 \longrightarrow 14 C_2H_4 + 12 CO_2 + 6 H_2O + 12 H_2 + CH_4$$

$$C_2H_6 \longrightarrow C_2H_4 + H_2$$
$$CO_2 + H_2 \longrightarrow CO + H_2O$$

FIGURE 9.9 Oxidative dehydrogenation of ethane with CO_2.

TABLE 9.6 Oxidative Dehydrogenation of Ethane to Ethylene Using CO_2

Catalyst	Conversion (%) CO_2	Conversion (%) C_2H_6	Selectivity (wt%) CH_4	Selectivity (wt%) C_2H_4	Selectivity (wt%) H_2/CO
Cr_2O_3/TiO_2	0.8	0.9	6.3	93.7	4.2
Cr_2O_3	16.8	12.1	4.6	95.4	4.8
Cr_2O_3/Al_2O_3	4.9	12.6	2.3	97.6	5.6
Cr_2O_3/ZrO_2	19.2	37.9	25.3	74.6	3.2
Cr_2O_3/SiO_2	9.6	38.8	4.2	95.7	1.4

3. The use of H_2 and O_2 mixtures is beginning to show promise as a replacement for H_2O_2, but safety issues will have to be seriously addressed.

4. Creative new catalytic technology can significantly reduce investment and carbon dioxide production through higher yields and few steps.

5. The direct use of CO_2 as a mild oxidant is an interesting new development that should be pursued aggressively.

6. Higher selectivity can also result in less processing, which is an unrecognized benefit that also reduces energy consumption and indirectly CO_2 emissions.

REFERENCES

1. Mars, P., and D.W. van Krevelen. *Spec. Suppl. Chem. Eng. Sci.* 3:41 (1954).
2. Sze, M.C., and A. P. Gelbum. *Hydrocarbon Processing.* 103-106 (February 1976).
3. Gaffney, A.M., A.C. Jones, J.J. Leonard, and J.A. Sofranko. *J. Catal.* 114(2):422-432 (1988).
4. Tremond, S.J., and A.N. Williamson. U.S. Patent 4,254,293 (1981).
5. Emig, G. and H. Hiltner. Oxidative coupling of isobutene in a two step process, pp. 593-602 in 3rd World Congress on Oxidation Catalysis. Elsevier Science B.V. (1997).
6. Contractor, R.M., M. W. Anderson, D. Campos, G. Hecquet, and R. Kotwica, WO 99/03809 (1999).
7. Haggin, J., *Chemical and Engineering News*, (April 3, 1995); Contractor, R.M., H.E. Bergna, H.S. Horowitz, C.M. Blackstone, B. Malone, C.C. Torardi, B. Griffith, U. Chowdhry and A.W. Sleight, *Catal. Today* 1(1-2):49-58. R.M. Contractor, H.E. Bergna, H.S. Horowitz, C.M. Blackstone, U. Chowdhry, and A.W. Sleight. *Stud. Sur. Sci. Catal.* 38 (Catalysis 1987):645-654 (1988).
8. Bergna, H. E., *ACS Syp. Ser.* 411:55-64. (1989) CODEN:ACSMC8 ISBN:0097-6156.
9. Kinoshita, H. and T. Ihara. (Mitsubishi Chemical Industries Ltd.), JP 98-294795.
10. Ushikubo, T., H. Nakamura, Y. Koyasu, and S. Wajki. (Misubishi Kasei Corp.), U.S. Patent 5,380,933 (1995); Harald, J., A. Tenten, S. Unverricht, and A. Heiko. (BASF) WO 9920590 (1999); Shinrin, T., T. Mamoru, and M. Ishii (Toa Gosei Chemical Industry Co., Ltd.). JP 09316023 (1997); M. Lin, and G. Buckley. (Rohm and Haas Ltd.), EP 0962253A2 (1999).
11. Okusako, A., T. Ui, and K. Nagai. (Sumitomo Chemical Co.) JP 95-171855 (1997).
12. Matsushita, T., T. Yamaguchi, S. Yamamatsu, and H. Okamoto (Asahi Chemical Industry Co., Ltd.), JP 10263399 (1998).
13. Mikami, Y., A. Takeda, and M. Okita, JP 09216850A2 (1997); Mikami, Y., A. Takeda, and M. Ohkita, DE 19734242A1 (1999).
14. Kharitonov, A.S., G.I. Panov, K.G. Ione, V.N. Romannikov, G.A. Sheveleva, L.A. Vostrikova, and V.I. Sobolev. U.S. Patent 5,110,995 (1992).
15. Hafele, M., A. Reitzmann, E. Klemm, and G. Emig. Pp. 847-856 in *3rd World Congress on Oxidation Catalysis*, Vol. 110, R.K. Grasselli, S.T. Oyama, A.M. Gaffney and J.JE. Lyons (Editors), Elsevier Science B.V. (1997).

16. Uriarte, A.K., M.A. Rodkin, M.J. Gross, A.S. Kharitonov, and G.I. Panov. Pp. 857-864 in *3rd World Congress on Oxidation Catalysis*, Vol. 110, R.K. Grasselli, S.T. Oyama, A.M. Gaffney and J.JE. Lyons (Editors), Elsevier Science B.V. (1997).
17. WO 9825698.
18. Clerici, M.G. Pp 21-23 in *Heterogeneous Catalysis and Fine Chemicals III*, M. Guisnet et al. (Editors), Elsevier Science Publishers B.V. (1993).
19. Gosser, L.W. U.S. Patent 4,832,938 (1989).
20. Hoelderich, W., German Patent DE 98-19845975.
21. Longya, X. , L. Liwu, W. Qingxia, Y. Li, W. Debao, and L. Weichen. *Natural Gas Conversion V*. Pp. 605-610 in *Studies in Surface Science and Catalysis*, Vol. 119, A. Parmaliana et al. (Editors), Elsevier Science B.V. (1998).
22. Qingxia, W., Z. Shurong, C. Guangya, et al., CN 87105054.4 (1987).
23. Wang, S., K. Murata, T. Hayakawa, S. Hamakawa, and K. Suzuki, *Applied Catalysis A: General* 196:1-8 (2000).

DISCUSSION

Dave Cole, Oak Ridge National Laboratory: I would like to ask two questions. First, would you comment on the discussion yesterday that pointed to the fact that mitigating CO_2 through commercial ventures such as this is a very small part of the CO_2 mitigation problem?

Second, as I recall, some years ago in one of the multilab reports on carbon management, there was a suggestion that one might generate carbon and hydrogen from methane in decarbonation-type reactions. Do you have anyone working on that? What might the catalytic and thermal requirements be to do that?

Leo Manzer: I agree with yesterday's discussion. For us, 500 million pounds a year of CO_2 is a lot of material, but still much smaller than carbon dioxide from electricity facilities. Large plants in the chemical industry generate a few hundred million pounds a year.

I can't deal with those other issues. We are working on oxidation catalysis, and we make a small improvement there. So it is a step in the right direction. I don't follow research in the conversion of methane into hydrogen and carbon. There has been work in that area, but the problem is that you still have to do something with the carbon. If you can use the carbon, fine. If it grows out of catalyst, the carbon will most likely coke up the catalyst; you would have to burn it off, and you would get CO_2 again. If you can use the carbon somewhere else, then this might be a way to make hydrogen.

Dave Cole: I have heard some discussion about actually trying to engineer the carbon to get it to hold onto the hydrogen in the form of either the NO_2 structures or even buckyballs. I don't know if anyone here has heard of this or done any work on it. If so, I would like to hear about it. In the recovery, is the problem that whatever surfaces you have get coated with the carbon?

Leo Manzer: Probably something like that is needed to generate low-cost buckyballs and buckytubes.

Chandrakant Panchal, Argonne National Laboratory: I think you can take more credit for CO_2 reduction. When you improve the selectivity of the chemical reactions, the energy cost for separation of products is reduced significantly. Most of the energy in the chemical industry goes to these operations. I think you can claim much more CO_2 reduction by improving selectivity. The Department of Energy's Office of Industrial Technologies program Vision 2020 has identified selectivity improvements as a way of reducing the need for separation—hence a way to reduce CO_2.

Leo Manzer: That is a very good point. I make this point in my report as well. If you have 100% selectivity, you can eliminate all refining and save a lot of energy that way.

Tobin Marks, Northwestern University: Let me follow up even more on that. If we are going into an era legislatively where there are punitive fines against the industry for CO_2 generation, then the things that you were talking about are going to be doubly valuable.

Leo Manzer: I think that is right, Tobin. I think this provides an opportunity for further research that may not be done as much in industry as it used to be. The opportunity is there for folks in academics to develop this technology. It takes a while, and this is tough chemistry, but it is an opportunity.

Tobin Marks: In Europe, is taxation levied against CO_2 producers? I have heard that it is in Norway. Can anyone tell us whether that is an issue in other countries?

David Keith, Carnegie Mellon University: There is no other systematic tax. Even the Norwegian tax is only on offshore emissions and is quite limited. There is a lot of talk and some targeted sectoral effects to put an effective price on carbon.

Leo Manzer: That is a good point. I think you have to do this globally to make it work.

John Frost, Michigan State University: I ask this question knowing that DuPont has a curious history in this area. One of the organisms that came out of the national labs—AFP-111, when grown on glucose, has yields in excess of 150% because it actively fixes CO_2 at atmospheric pressure during the fermentation. In what potential scenario could you see this type of process becoming competitive with what you are practicing now?

Leo Manzer: You may be aware that we have a fermentation process that we are developing to take glucose to 1,3-propane diol, which is in the pilot plant stage right now.

This is an example of how a biochemical route, starting with something like glucose, can be cost-effective. It can be even better than chemical routes to make propane diol. Does that answer your question? We are not starting with methane.

John Frost: I think it is a question of issues of displacement technology. Where is the crossover point in this game? You have a facility that is obviously well established.

Leo Manzer: Replacement economics are very tough, almost impossible. It is just the game we are playing these days. This is why it is nice to work in an area where there is decent growth, so that every few years or so when you have to build another plant, you can include any new technology in it.

Dave Thomas, BP Amoco: As was pointed out yesterday, the amount of carbon dioxide produced by the chemical industry is relatively small compared to many of the other processes. Every time you take something up and boil it and then cool it off through some mechanical or chemical process, you consume a great deal of energy.

I think the contribution that the chemical industry can make in the area of CO_2 mitigation is in high-efficiency processes that reduce the amount of heat input and enhance the energy consumed. As you point out, highly selective reactions could be very valuable.

As an aside, one of the people I talked to in our chemical business said that if we have to go to full-scale sequestration from chemical plant CO_2 effluents, this would essentially cause its margins to evaporate. In other words, the business would not be economically viable unless everybody else had to do the same and the price was just raised.

The cost of CO_2 mitigation for the chemical industry is not small. I was struck by your earlier comments about some of the oxidative reactions using air. You pointed out that in reaction with butane, you have to avoid the explosive limit, and you can produce a lot of CO_2 by accident if you don't avoid the combustion limits. Have you thought about using carbon dioxide-oxygen mixtures rather than air in these kinds of reactions?

Leo Manzer: CO_2 often is recycled in some of these processes. It is not specifically mixing CO_2 and oxygen. That will change the flammability envelope significantly, I think. We have not done this. It may be an opportunity for somebody to look at.

Olaf Walter, Forschungszentrum Karlsruhe (Germany): Oxidations are always a target for increasing selectivity. In recent years, there also have been investigations using microreactors for heterogeneous oxidations, which may increase selectivity, often by avoiding hot spots on the catalysts.

A second technique, also in the same direction, might be the use of supercooling carbon dioxide as a solvent, where you have a one-phase system in the oxidation. This is then a homogeneous reaction, and a catalyst may not be needed. I think we should point out that there is the potential of still further optimization of these processes.

Richard Alkire, University of Illinois: It is very exciting to hear of the need for new chemistry and also new processing. In chemical engineering at least, that has dropped away from the academic side over the years. In my opinion, in the earlier part of the century, chemical engineers in academics actually participated in the invention of new processes. When I think of the deemphasis over the years on the academic side coupled with the reduced industrial research investment, it seems to me that there is the potential for a serious disconnect which may already exist. If I think of trying to do chemical engineering as running across the street at any other university, I can imagine having difficulty engaging in conversation that might be helpful in the way you suggest is needed.

Yet you have mentioned certain research areas in which academics could make contributions. Could you say exactly what the interesting chemistry or engineering is that is needed here? Also, do you have any thoughts on how to engage that part in a useful way?

Leo Manzer: First, you have to have funding available in both chemistry and chemical engineering so that you have enough people available to discuss things with each other. This is the most important part of the process. Consider DuPont's work on the oxidation of butane to maleic anhydride. We worked on this for 15 years. In that situation, chemistry and chemical engineering were completely integrated with folks constantly talking together.

The engineering aspects of running an oxidation reactor in an anaerobic mode like that with lots of butane around, was very complicated. This is one of the biggest issues. If people actually do look at developing some of these other processes, it is scale-up that will be most difficult. The scale issues are incredibly complicated. There is a real opportunity where the funding is available for engineers and chemists to get together to work jointly on these projects. I don't know how to make this happen except to have money available to support your research programs and those of others. I mean, without the people there to do the research, it will never happen.

10

Increasing Efficiencies for Hydrocarbon Activation

Harold H. Kung
Northwestern University

Hydrocarbons derived from fossil fuel are the main source of energy and raw material for petrochemicals in the industrial world. When not used in combustion to generate power and heat, fossil fuels are refined in various petrochemical transformation processes into purer and higher-valued products. This chapter continues the discussion by Leo Manzer to address opportunities for research in chemical sciences to reduce carbon (dioxide) emission. Although the large majority of carbon emission is from power generation and transportation, the discussion here focuses on hydrocarbon conversion in the chemical processing industry, with only a brief discussion of hydrocarbon conversion in fuel cell applications.

There are different ways to increase hydrocarbon conversion efficiency to minimize carbon oxides emission. In a chemical transformation process to form a desired product, carbon oxides can be emitted as a by-product in the chemical reaction, as a result of the generation of power needed to effect the desired chemical transformation, or in the generation of reactants needed for the reaction. Increased efficiencies in any of these aspects would reduce carbon emission. In this chapter, examples are presented to illustrate research opportunities that could reduce (1) carbon-containing wastes, (2) hydrogen consumption or wastes, and (3) energy needed for chemical transformation.

REDUCING CARBON-CONTAINING WASTES

In general, chemical transformation processes can be exothermic or endothermic. The examples of catalytic selective oxidation presented by Leo Manzer in Chapter 9 are exothermic reactions, in which carbon oxides are formed as by-products. Increasing selectivity for the desired product would lower carbon emission because of less CO_x formation in the reaction and lower energy consumption needs for downstream separation and purification. One example mentioned in Chapter 9 is the production of maleic anhydride by selective oxidation of butane (Equation 10.1). The commercial yield is only about 50%. That is, about 4 moles of CO_x are formed as reaction by-products for every mole of maleic anhydride produced. There is substantial room for improvement:

$$C_4H_{10} + 7/2\ O_2 \rightarrow C_4H_2O_3 + 4\ H_2O. \qquad (10.1)$$

Another example is the catalytic oxidative dehydrogenation of propane to propene. This is not yet a commercial process because of the low yield of propene using known catalysts. Figure 10.1 summarizes the reported selectivity as a function of conversion of propane for the better catalysts. The data show that selectivities approaching 100% can be obtained only at low conversions, suggesting that the low selectivities at high conversions are due to the faster oxidation of propene to CO_x than of propane to propene. The former reaction is faster because allylic C–H bonds in propene are much weaker and more reactive than the C–H bonds in propane. Consequently, if catalytically active sites that are indifferent to C–H bond strengths can be constructed, such that the catalyst promotes the reaction between oxygen and propene as fast as that between oxygen and propane, then the selectivity-conversion relationship shown by the solid line in the Figure 10.1 could be obtained. Unfortunately, at present, there is insufficient information on the nature of the active sites in mixed oxide catalysts reported for this reaction to permit designing such an active site. One difficulty in attempts to elucidate the nature of active sites in mixed oxide catalysts for selective oxidation in general is the poorly crystalline state of the solid at the active site because of the facile motion of oxygen ions under reaction conditions. Development of more informative experimental methods and computational techniques will be very valuable in this area.

In addition to exothermic reactions, there are many chemical transformation processes that are endothermic. Examples include steam reforming of methane to generate hydrogen and cracking of hydrocarbon in the fluid catalytic cracking process. For these reactions, heat is required, which is often

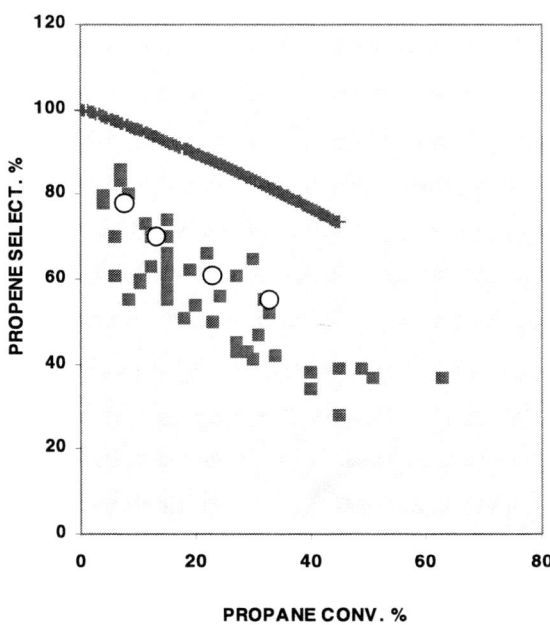

FIGURE 10.1 A summary of reported selectivity-conversion relationships in the oxidative dehydrogenation of propane to propene. Solid line is the relationship for a sequential reaction of propane to propene to CO_x if the first-order reaction constants for the two steps are identical. Open circles are data for homogeneous noncatalytic reaction.

supplied by the combustion of carbon-containing fuels. In these cases, it may be possible to use unwanted by-products of the reaction as fuel, and the most efficient process (in terms of total carbon emission) would be one in which the fuel value of the waste by-products matches the endothermicity of the reaction.

REDUCING HYDROGEN CONSUMPTION AND WASTES

The hydrogen-to-carbon ratio in most petrochemicals is higher than in crude oil. Therefore, hydrogen must be added in their production. Industrial production of hydrogen is mostly by the energy-intensive steam reforming of methane or, less frequently, by partial oxidation of methane. Both processes emit carbon oxides. Thus, reducing hydrogen consumption in a process would reduce carbon emission.

For example, the production of butanediol from butane involves three steps (Figure 10.2 solid arrows): (1) partial oxidation of butane to maleic anhydride, (2) selective hydrogenation of maleic anhydride to tetrahydrofuran, and (3) hydration of tetrahydrofuran to butanediol. From the stoichiometry of the reactions, for every mole of butanediol produced—even if each step proceeds with 100% yield—3 moles of hydrogen are consumed.

In principle, it should be possible to produce the same product by selective oxidation of butane with oxygen (Equation 10.2 and Figure 10.2, dashed arrow).

$$C_4H_{10} + O_2 \rightarrow HOCH_2CH_2CH_2CH_2OH. \qquad (10.2)$$

The feasibility of such direct oxidation of a terminal carbon, has been demonstrated recently in the selective oxidation of hexane to adipic acid (1,6-diacid)[1] by molecular oxygen catalyzed by cobalt-containing aluminum phosphate molecular sieves. Molecular modeling suggests that the hexane molecule is adsorbed in the pores of the molecular sieve in such a configuration that the two end carbons are close to the cobalt ions in the framework, enabling selective activation of the terminal C–H bonds. This parallels enzymatic action, where specific binding of the substrate with the protein leads to a favorable configuration for selective reaction with the active center.

There are other processes in which potential exists to reduce hydrogen consumption. For example, in the ammoxidation of propene to acrylonitrile (Equation 10.3), ammonia produced from hydrogen and nitrogen is used:

$$C_3H_6 + 3/2 O_2 + NH_3 \rightarrow CH_2CHCN + 3H_2O. \qquad (10.3)$$

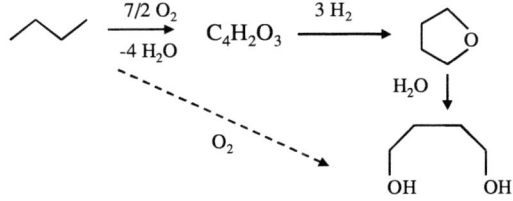

FIGURE 10.2 Reaction pathways for the conversion of butane to butanediol.

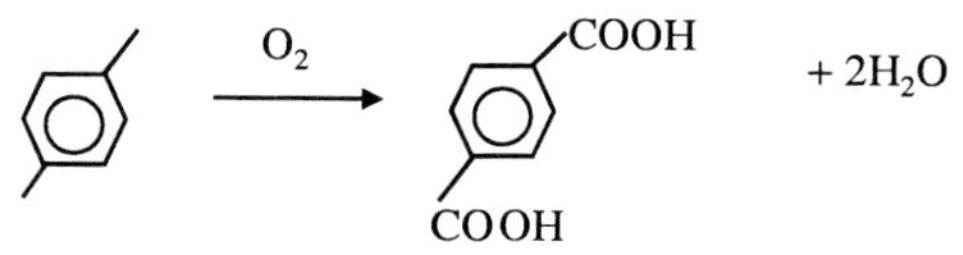

FIGURE 10.3 Current method for the formation of terephtalic acid.

If a new method can be found that uses N_2 instead of NH_3, there will be less carbon emission in the overall production process.

Another possibility to reduce hydrogen consumption is in the production of *p*-terephthalic acid, which is currently achieved by oxidation of *p*-xylene. Two moles of water are formed for each mole of terephthalic acid formed (Figure 10.3). In principle, the product can be produced from benzene without loss of hydrogen (Figure 10.4). In order to achieve this, new catalytic processes must be developed for selective hydroxylation of benzene and carbonylation of catechol. Currently, selective hydroxylation can be accomplished using a titanium silicalite catalyst and peroxide as oxidant, but a method using O_2 is needed. Carbonylation of alcohol is known. The carbonylation of methanol to acetic acid using rhodium complex catalysts is a commercial process, although—as discussed next—improvement is possible.

REDUCING ENERGY NEED

In addition to the energy needed for endothermic reactions, there is substantial energy consumption in the various separation and purification steps in a chemical transformation process. Increasing selectivity in chemical reactions of the desired product, thereby decreasing the amount of by-product that has to be separated, would reduce the energy need accordingly. Additional opportunities arise when processing conditions can be modified, especially when radically new processes can be discovered. One example is methanol carbonylation.

Recently, the Cativa process for liquid-phase carbonylation to produce acetic acid has been commercialized.[2] This process uses an iridium-based catalyst instead of rhodium, produces less propionic acid and acetaldehyde, and uses a much lower water concentration in the reaction mixture. The last aspect results in a reaction product stream that contains less water, so much less energy is needed for distillation to separate water from acetic acid.

There are opportunities for further improvement. In principle, the conversion of methanol to acetic acid requires only the insertion of a CO molecule (Equation 10.4). If this could be achieved in the vapor phase, the need to separate acetic acid from water would be eliminated:

$$CH_3OH + CO \rightarrow CH_3COOH. \tag{10.4}$$

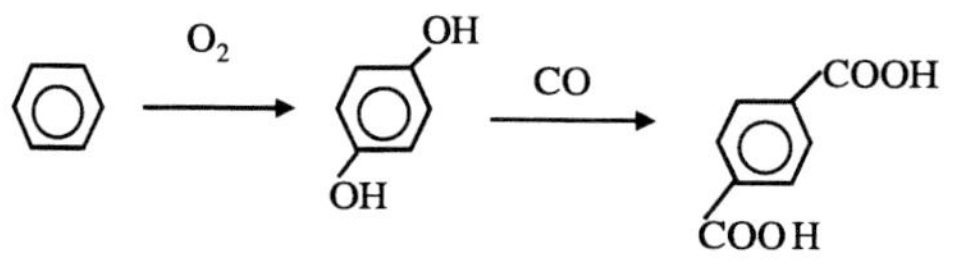

FIGURE 10.4 Preferred method for the formation of terephtalic acid.

Vapor-phase carbonylation catalyzed by copper-MOR (Cu-mordenite) has been reported.[3] However, a successful process will require a better catalyst system. The Cu-MOR catalyst suffers from deactivation, and there are substantial side reactions of hydrocarbon formation over the Brønsted acid sites.

At present, many starting materials for chemicals are light hydrocarbons obtained from petroleum fractions. As the petroleum refining process improves and the demand for chemicals increases, it might become necessary to develop specific processes to produce raw materials on demand. At that time, there would be a need to synthesize specific hydrocarbons to improve the flexibility of operation. For example, the recent report of selective dimerization of alkene could enable efficient synthesis of medium-length hydrocarbons without the need for further separation.[4] It should be equally useful to be able to selectively cleave a longer hydrocarbon molecule into shorter ones instead of the current nonselective cracking. The potential to improve process efficiency could be further enhanced if selective isomerization of hydrocarbon could be realized. This includes both skeletal and double-bond isomerizations.

The examples mentioned above require development of new chemical transformation processes. Associated with these processes is the need to develop new catalytic systems. Improved understanding of current catalytic systems would facilitate the scientific discovery of new ones. New experimental and computational techniques are needed. Insufficient understanding of catalysis in general is demonstrated by the system of supported gold catalysts: Although these catalysts have a wide variation of reaction selectivity, there is little understanding of the nature of their active sites.

HYDROCARBON ACTIVATION FOR FUEL CELLS

Fuel cells are rather efficient energy conversion devices. Currently, the transportation industry is seriously investigating the use of fuel cells as the power source for automobiles to replace the internal combustion engine. For operational reasons, proton-exchange membrane fuel cells, combined with on-board reforming of a liquid fuel to supply hydrogen, is considered the most feasible candidate for the near future. Many technological advances, such as improvements in the fuel processing unit, are needed before fuel cell-powered vehicles are ready for large-scale market penetration.

The current fuel processing unit consists of a fuel reformer that catalytically converts the liquid fuel into a mixture of hydrogen, CO, CO_2, and water. The CO concentration is then reduced using a water-gas shift unit. Much research is devoted to improve the activities of the catalysts in both the reformer and the water-gas shift unit to reduce their weight and volume. In the autothermal mode of operation of the reformer, there is first complete combustion of the fuel into water and CO_2 until nearly all of the feed oxygen is consumed, followed by steam reforming of the fuel to generate hydrogen. Because steam reforming is relatively slow, the bulk of the reformer unit is occupied by catalyst for steam reforming. This provides opportunities for new approaches to fuel reforming. One possibility is to effect fuel conversion to hydrogen by selective oxidation at low temperatures. The low temperature could lead to lower CO concentration because of the more favorable equilibrium dictated by the water-gas shift reaction. If the slow steam reforming reaction is no longer needed, the size of the reformer unit could be reduced substantially.

SUMMARY

This chapter provides a few examples of the many opportunities that exist to reduce carbon emission in hydrocarbon conversion processes. Potentially substantial reduction could be achieved by new, innovative catalytic systems. In addition to improving selectivity in chemical reactions, new processes that match the number of carbon and hydrogen atoms in the reactants with those in the products should be developed. These processes would reduce both carbon and hydrogen wastes. The latter is important because there is carbon emission in hydrogen generation.

In order to facilitate discovery of revolutionary catalytic systems, there needs to be better understanding of the interaction of the active sites of a catalyst with various bonds in a reactant molecule and in the product. There is also a need to learn to beneficially use the interaction of the nonreacting portion of a reactant molecule with atoms in the catalyst away from the active site as a means of controlling the configuration of the reactant molecule at the active site and, thus, the reaction selectivity. These goals can be accomplished by developing new insitu characterization techniques and more powerful computational methods to understand existing catalytic systems better.

REFERENCES

1. Raja, R., G. Sankar, and J.M. Thomas, 2000. *Angew. Chem. Int. Ed.* 39:2313.
2. Sunley, G.J., and D.J. Watson, 2000. *Catal. Today* 58:293.
3. Ellis, B., M.J. Howard, R.W. Joyner, K.N. Reddy, M.B. Padley, and W.J. Smith, 1996. *Proc. 11th Intern. Cong. Catal, Stud. Surf. Sci. Catal.* 101;771.
4. J. Christoffers and R.G. Bergman, 1998. *Inorg. Chim. Acta.* 270:20.

DISCUSSION

David Bonner, Rohm and Haas Company: I think your talk and Dr. Manzer's talk together provide a beautiful synergy, and it is great that you all collaborated. If you try to boil the messages from both of the talks down to the main concepts, I would say it is clear that although there is a broad array of possibilities available to us, reinvestment economics limit some of the things that can be taken through practical development.

If reinvestment economics do shift because of carbon management considerations, through either government regulation or economics of another sort, what areas of research funding, in your opinion, should be emphasized more than they are now by policy makers in the federal government?

Harold Kung: In my summary, I suggest a number of things. We need to better understand the interactions of the entire molecule with the catalysts, not just the active sites, and how to manipulate them. Sir John Thomas's example makes use of the structure of molecules. Once we know what is required for a particular reaction, we can design a catalyst that involves how the whole molecule interacts and how the active site acts on the portion of the molecule that we want to change. I think there are a lot of opportunities there. For example, Davis and colleagues has been trying to synthesize a cavity based on what they want in their reaction.

That is just the beginning. There are a lot of research opportunities in so-called nanotechnology for catalyst design beyond the immediate active site. In my opinion, this is an area that will be very fruitful; the preceding example suggests that it can be done.

Hans Friedericy, Honeywell International: To come up with a fuel cell car that really is long-lived and will eventually compete with a diesel car, one of the keys is the membrane itself and the catalyst inside. Developing a less expensive catalyst for fuel cell systems is an area that is lacking sufficient research. You mentioned fuel cells—are you involved in that area?

Harold Kung: I mostly confined my few comments on the fuel cells to the hydrocarbon activation aspect, so I am talking about the hydrogen fuel cell. A lot of progress has been made in reducing platinum usage in the anode. I think James Spearot mentioned that yesterday. There is also a lot of work on increasing the temperature tolerance of the membrane.

Hans Friedericy: Has anyone been successful?

Harold Kung: There has been substantial progress, but further improvements are needed for commercialization.

Hans Friedericy: The reformers that go with fuel cell engines need a lot more work too. Honeywell has studied at least 25 different areas, and we yet have to find a membrane that can take higher temperature. Even if we have a membrane that has a better catalyst in it, we are still stuck with a fuel cell that is too expensive.

Harold Kung: Reducing the cost is definitely an issue. Many companies are working on cost issues. Fuel cell research has received a higher level of attention than ever before. It seems premature to conclude that the current problems cannot be overcome. In my presentation, I suggested a totally different way of fuel reforming. Rather than just thinking about how to come up with a better, more actively reforming catalyst in the traditional way, let's look at the process totally differently. Perhaps there can be completely new proton transfer membranes as well.

There are research opportunities in what I would call pre-competitive research that is very suitable for government funding.

Hans Friedericy: I also think you can get a lot of funding for it.

Harold Kung: Yes, fuel cells are receiving more attention these days.

Klaus Lackner, Los Alamos National Laboratory: I would like to point out that one of the largest sources of electricity is ultimately coal-burning power plants. If you really want to get the effluent clean and want to capture your CO_2, ideally, in a separate stream, you end up going through a calcification process. The issue of catalysts and how to make this work is certainly on the mind of everyone working in that area. I think that if the power plant of the future is a coal plant, it is likely to be a gasification plant because I see this as the only way of collecting CO_2 and all the other pollutants for that matter.

Tobin Marks, Northwestern University: This issue concerns available funding for topic A or B. Funding for academic research in the catalytic area in the United States is not very large, and the funding that is there has been shrinking. I think the program managers here would verify that. If we want to attack some of these problems, I think we have to generate more interest nationally as well as generate arguments for why this funding really is needed.

11

Commodity Polymers from Renewable Resources: Polylactic Acid

Patrick R. Gruber
Cargill Dow LLC

Of all of the potential products made from annually renewable resources, polylactic acid (PLA) is currently the most interesting. PLA has the properties and value to create a viable business opportunity that will enable Cargill Dow to create new market opportunities. These new markets value material properties, as well as environmental attributes.

In business, perceptions of commercial opportunity drive investment. In research, I believe the same to be true. Cargill Dow LLC is building a world-scale manufacturing facility for PLA with a capacity of 300 million pounds a year at a cost of several hundred million dollars. This investment was made because (1) the product performance is adequate and (2) the price delivered to the marketplace is reasonable. The fortunate consequence is that because of these two factors, customers are lining up to buy the PLA.[1] We had a very strong opinion that our products must provide value as a material. As a result, we invested tens of millions of dollars and focused our R&D to discover this value. While it is true that our products have interesting environmental benefits, these benefits alone do not drive investment in either research or manufacturing facilities. There is an important principle here. Investment to solve CO_2 in the atmosphere ought to focus on products that add value to society—meaning products that work well, are competitively priced, and address the CO_2 problem. PLA provides the first real-world example that does all of these things in this new market of renewable resource-based chemical products.

However, many questions are being asked about products made from renewable resources. Can products with the performance of conventional thermoplastics actually be made economically? Do these materials really have the performance required? Are these products really more environmentally friendly and how do you know? What are the issues? This chapter addresses those questions and draws generalizations that can be applied to other chemical products made from renewable resources.

MANUFACTURING OF PLA

The raw material for PLA manufacturing is any fermentable sugar. The cheapest and most abundant source of sugar in the world is currently corn. Hence, Cargill Dow LLC (CD) is planning to start with corn. In the long run, I expect that as technologies geared toward biomass-to-sugar conversion mature, CD will use these technologies to generate fermentable feedstocks. Figures 11.1 and 11.2 outline the manufacturing steps of PLA. Carbon dioxide is fixed in crops to make starches. Agriprocessing businesses like corn wet-milling convert the starches to simple sugars. CD buys the sugar and uses it to ferment lactic acid. Using chemical processing techniques, lactic acid is converted efficiently to lactide, a ring-form dimer of lactic acid. Lactides are excellent polymerization starting materials because they are reactive and anhydrous, and they polymerize in the melt. PLA is made through ring-opening melt polymerization. The overall process is sufficiently efficient in terms of yield and energy that the products are economically viable.

Market Opportunities

PLA currently finds demand in three market areas: fibers, packaging, and chemical products. Significant research investment in product development has revealed product attributes that are valuable and the knowledge of how to best to use them in the marketplace. Fibers and packaging provide the strongest examples of how PLA attributes bring value to a market.

PLA fibers combine the comfort and feel of natural fibers with the performance of synthetics (Table 11.1). The unique property spectrum of PLA fibers allows the creation of products with superior hand and touch, drape, comfort, moisture management, ultraviolet (UV) resistance, and resilience. Combining these performance features with the features of natural fibers enables PLA to be used in a wide spectrum of products including apparel, carpet, nonwoven fiberfill, and household and industrial markets (Figure 11.3).

PLA apparel, carpets, and nonwovens are already in test market. Consumers' reports indicate that the products actually work well, and they appreciate the products being made from renewable resources. Of course when consumers indicate that they appreciate a product made from renewable resources, they expect that there should be some measurable advantage regarding the environment compared to traditional petroleum-based products.

PLA Packaging

PLA polymers for packaging applications exhibit a balance of performance properties that are comparable and in certain cases superior, to traditional thermoplastics. PLA is useful in coated paper, films, rigid containers, bottles, and a variety of other packaging applications (Figure 11.4). However, there are two specific packaging areas that have received initial focus—high-value films and rigid thermoformed containers. Functional properties and their benefits are listed in Table 11.2.

The combination of functional properties provides the commercial drive for PLA. A close look at the properties listed in Table 11.2 reveals that the technical attributes primarily benefit manufacturers and converters. The exceptions, renewable resources and compostability, are end-user and consumer-oriented attributes.

The market development for packaging is quite different than for fibers. PLA fibers benefit the consumer directly. For example, not only are the products—a shirt, for example—more comfortable (I can detect it myself as a consumer), they are also made starting from a natural product (a perception). So in fibers, the combination of direct consumer benefit and easily communicated perceptions helps to drive the potential of PLA.

In packaging market segments, consumers' concern for the environment has driven manufacturers to want to adopt new technologies. Led by Europe and Japan where environmental concerns receive a

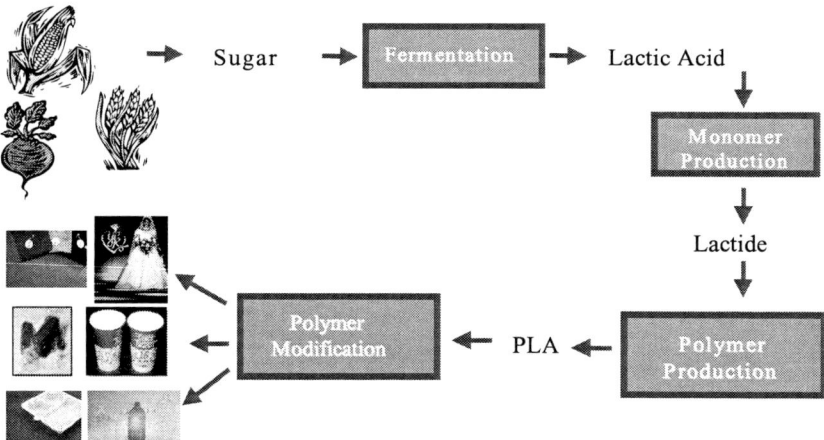

FIGURE 11.1 Overview of manufacturing processes for PLA.

higher priority than in the United States, converters and manufacturers are actively developing packaging products with improved environmental performance.

In contrast to fibers, with packaging market segments consumers will probably not directly detect many of the technical attributes and benefits. Although it is true that PLA can make a better package, consumers don't buy packages. They buy the products in the packages. We find, however, that consumers expect that if technology exists to make a package more "environmentally friendly," companies should use it—as long as it doesn't increase the price of the product too much. This market insight is critical when investing in technology aimed at environmental attributes. Being environmentally friendly is worthwhile, but only at a certain cost, and it must provide clear benefit.

I believe this insight should be extrapolated to the type of research investment necessary to resolve environmental problems. The answers to the questions, What is the measurable benefit to consumers? and How much should be spent? depend on the benefit provided. Research for CO_2 in the atmosphere

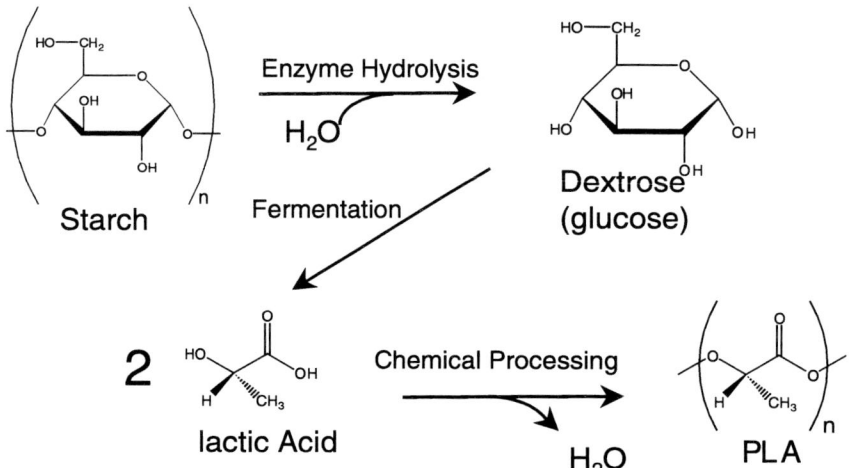

FIGURE 11.2 The processing route to PLA combines bioprocessing and chemical processing

TABLE 11.1 Summary of Fiber Properties

Fiber Property	Synthetics			PLA	Natural Fibers			
	Nylon 6	PET	Acrylics	PLA	Rayon	Cotton	Silk	Wool
Specific gravity	1.14	1.39	1.18	1.25	1.52	1.52	1.34	1.31
Tenacity (g/d)	5.5	6.0	4.0	6.0	2.5	4.0	4.0	1.6
Moisture regain (%)	4.1	0.2-0.4	1.0-2.0	0.4-0.6	11	7.5	10	14-18
Elastic recovery (5% strain)	89	65	50	93	32	52	52	69
Flammability	Medium	High smoke	Medium	Low smoke	Burns	Burns	Burns	Burns slowly
UV resistance	Poor	Fair	Excellent	Excellent	Poor	Fair-poor	Fair-poor	Fair
Wicking (L-W slope; higher slope, more wicking)	—	0.7-0.8 (no finish)	—	6.3-7.5 (no finish) 19-26 (with finish)	—		—	

NOTE: g/d=grams per denier; L-W=Lucas-Washburn Equations; PET=poly(ethyleneterphalate).

should take a product approach. Products must provide value to consumers. The product could be a reforested desert wasteland or it could be PLA—these are tangible to the people who pay—the consumers.

Market Potential

In the long term, PLA can compete successfully in several markets with an annual volume of more than 6.6 billion pounds. With technology improvements in manufacturing and processing, these markets

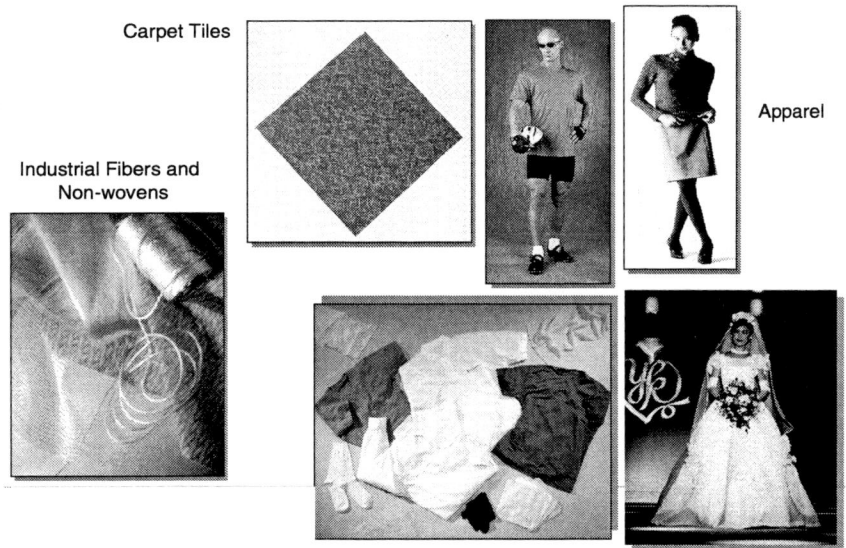

FIGURE 11.3 Typical PLA fiber applications.

FIGURE 11.4 Examples of PLA packaging applications.

could expand to about 10 billion pounds of PLA annually. The market value of annually renewable resource-based thermoplastics from PLA would be at least $6 billion to $10 billion per year.

In addition, lactic acid can serve as a chemical intermediate. As our scale increases and the costs of the lactic acid manufacturing process are reduced, we expect that lactic acid will be inexpensive enough to enable several other end markets in the chemical industry. This concept is illustrated in Figure 11.5, which shows the wide variety of chemicals other than PLA that could be made from lactic acid. These chemicals add an additional 3 billion to 4 billion pounds and market value of $1 billion to $4 billion per year to the estimated PLA value.

Why didn't CD initially target traditional chemicals based on renewable resources? The answer is simple: we couldn't convince ourselves that the research investment was justified because the target markets were not big enough. We could not create a hypothetical process that would produce lactic acid cheaply enough to create the profit necessary to justify the research funding. We instead developed a

TABLE 11.2 PLA Functional Properties for Packaging

Functional Property	Packaging Improvement
Deadfold, twist, and crimp	Improved folding and sealing
High gloss & clarity	Package aesthetics
Barrier properties	Grease and oil resistance
Renewable resource	Made from CO_2
Flavor and aroma	Reduced taste and odor issues
Low-temperature heat seal	Stronger seals at lower temperatures
High tensile and modulus	Wet paper strength, ability to down-gauge
Low COF, polarity	Printability
GRAS status	Food contact approved
Compostable	Compostable, low "green" tax

NOTE: COF=Coefficient of Friction; GRAS=Generally Recognized As Safe.

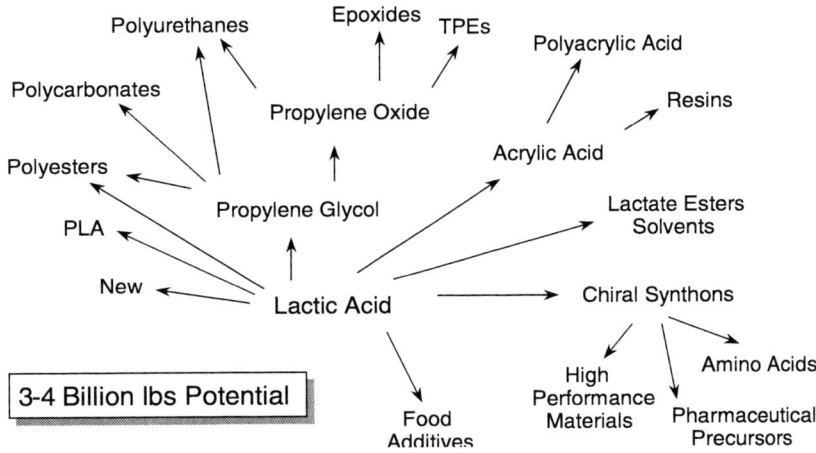

FIGURE 11.5 Chemicals that could be made from lactic acid.

new product (PLA) in a value-added, large market category that, if it succeeds, allows scale and scope economies. This approach combines more risk, more reward, more complicated market development, and more rapid advancement of our knowledge.

The commodity chemical market dictates that value is created when performance is equal at lower cost. When faced with a project choice where the target molecule is an exact replacement of a commodity, I look for technology that provides at least a 50% cost reduction or the project is not worth pursuing. I've reviewed many proposals where process research work is being justified because it is more environmentally friendly—usually meaning lower energy and/or CO_2 emissions—but at a process economic cost close to the incumbent commodity product. This strikes me as underachieving. With modern biotechnology, bioprocessing, and chemical processing technology, more focus should be on creating environmentally friendly processes *and* 50% cost reductions. The lesson is that just because something is made from renewable resources does not make it better. It *must* provide value in the marketplace.

PLA takes advantage of a biological system to do chemistry that traditional chemical techniques can't. Fermentation of sugar to lactic acid does something that we cannot do by chemistry, namely, produce chiral lactic acid in high yield. Figure 11.6 shows the optical isomers of lactic acid.

Not only do bioprocessing and fermentation provide chiral lactic acid, they do so inexpensively. Control of chirality allows us to change the polymer performance by changing the optical activity of the lactic acid units in the polymer backbone. The result is a family of products made from lactic acid, with properties that can reach the wide range of applications previously discussed.

The two isomers of lactic acid can give rise to three lactides: l-, d-, and *meso*. The fermentation process makes *l*-lactic acid exclusively. The chemical processing steps allow us to racemize small amounts of l-lactic acid to d-lactic acid. This provides us with three dimers for polymerization. Figures 11.7 and 11.8 are cartoon representations of the polymer structures we can achieve from the three lactides. By varying the amount and sequence of *d*-lactic units in the polymer backbone, we can change product properties, such as melt behavior, thermal properties, barrier properties, and ductility.[2]

Our market opportunity results from the combination of fermentation, bioprocessing, and conventional chemical processing. The optical specificity of the biological system provides value that cannot yet be achieved through chemical processing. The chemical processing steps allow us to specifically combine the chiral lactic acid units to make valuable materials. The combination of fermentation or

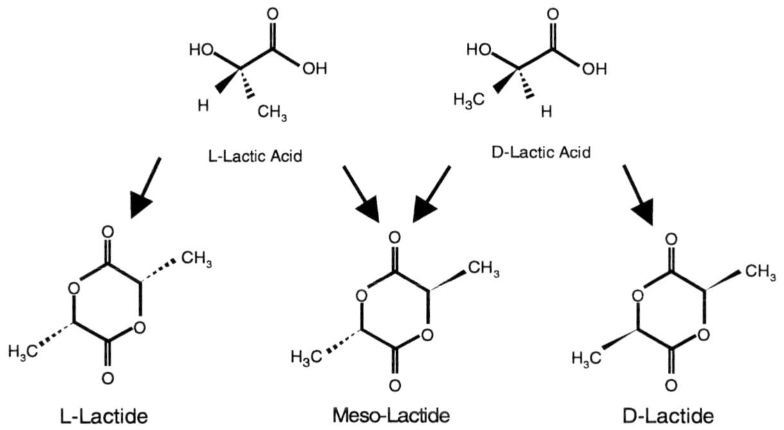

FIGURE 11.6 The optical isomers of lactic acid.

bioprocessing and chemical processing provides specific value that can't be achieved alone. Using the chirality of the molecules, we are able to create a wider range of performance in a wider range of products. Our market opportunity is therefore bigger. I believe that this insight is general and would extend to other products made from renewable resources.

Renewable Resources, CO_2, and PLA

Whenever someone hears that PLA is made from renewable resources, there is an immediate expectation that it must be better for the environment. Two questions soon follow: (1) How does one know it is better? and (2) Better than what? Energy and CO_2 are the focus of this discussion. PLA provides an example of what is required to answer these questions successfully.

All of the carbon in PLA comes from carbon dioxide in the atmosphere. How? All of the carbon in lactic acid comes from glucose, which is made by plants via photosynthesis. However, like all manufacturing processes, the production of PLA requires energy. The question is, Does the energy required to drive the processing cause more carbon dioxide emissions than the amount of carbon dioxide fixed in PLA? This question can be addressed using a complete "cradle-to-grave" life-cycle inventory (LCI).

LCIs were designed to provide an accounting of all of the inputs and outputs across the whole business system. LCIs should be used internally by companies to gain insight into where to effectively

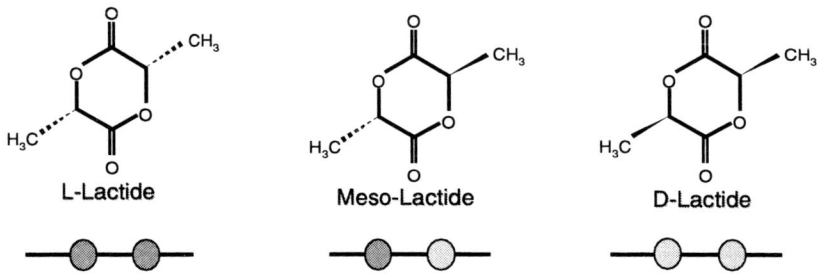

FIGURE 11.7 Possible lactide structures. The dark color represents an l-lactic acid residue. The light color represents a d-lactic acid residue.

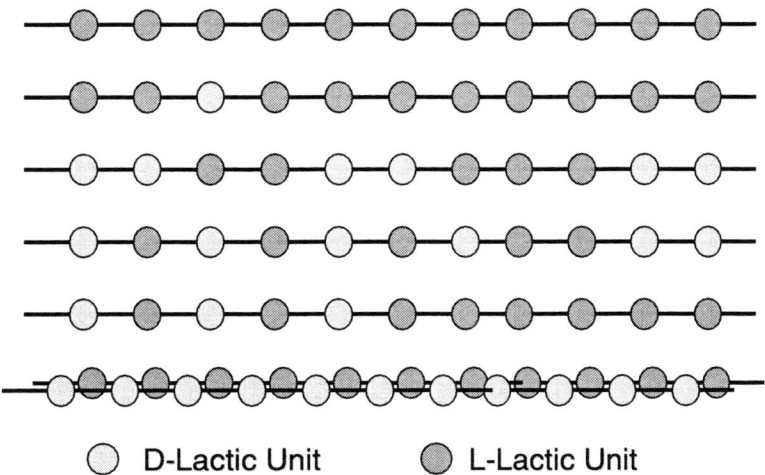

○ D-Lactic Unit   ● L-Lactic Unit

FIGURE 11.8 Examples of PLA structures. The dark color represents an l-lactic acid residue. The light color represents a d-lactic acid residue.

improve their processes by reducing energy and emissions. LCI also causes companies like ours to take a very keen interest in what our suppliers and customers do—they are part of the business system.

The business system for thermoplastics is illustrated in Figure 11.9, where both a renewable-based system and a petrochemical-based system are shown. The major difference in these two systems is that in the renewable-based system the raw material is corn, while in the petroleum-based system it is petroleum. Both systems use fossil resources for process energy. Both systems make a plastic product that needs to be disposed at the end of its useful life. Another difference between a PLA and a petrochemical-based system is that PLA can be conventionally recycled, re-monomerized back to lactic acid, or composted.

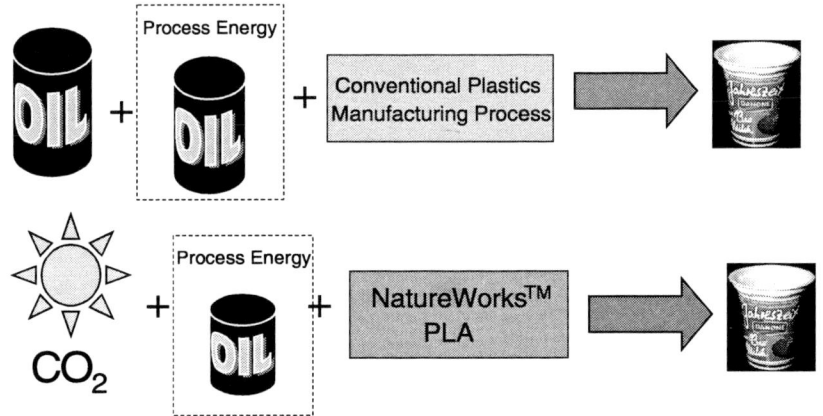

FIGURE 11.9 The business system for thermoplastics. With renewable-resource based products such as PLA all of the mass of the polymer originates from carbon dioxide. Conventional plastic materials use fossil resources. Both petrochemical-based plastics and PLA require process energy. The expectation is that the overall amount of fossil resources is less for PLA compared to petrochemical based products.

Each time an LCI is put forth, it needs to be specific for products and end fate to make valid comparisons and interpret the cradle-to-grave energy use and emissions.

LCI techniques are problematic because the data for raw materials are often not specific enough and cradle-to-grave systems are complicated. Raw data for inputs tends to be historical rather than forward looking. Agricultural data are frequently old, and the sample sizes often represent large regions. In the future, as we begin purchasing sugar from Cargill in Blair, Nebraska, we need to understand the agricultural practices used to grow corn. However, agricultural practices are changing rapidly and the data available aren't current or specific enough to represent the agricultural practices around Blair. What does this mean for PLA's LCI? We'll have to collect data and project it.

We have analyzed Cargill's wet-milling operation as well as our lactic acid and PLA processes and converted the data for use in our LCI. Our raw data are proprietary because they use real engineering data wherever possible. Figure 11.10 shows all of the inputs and outputs that are considered for an LCI of PLA. We have attributed all of the agriculture inputs to dextrose in our analysis to avoid debate over carbon dioxide capture by crop residues left on the field. We have taken into consideration all of the transportation to ship our products around the world. We have even taken into account the manufacturing of the agricultural machinery (it turns out not to be significant to the analysis).

Another complication in LCI comes from the time horizon considered in the business system. For example, in most developed countries of the world the fate of plastic material is likely to be incineration. Whether the plastic is a "disposable" or a "permanent" product, if it makes it to a waste disposal system, it is most likely incinerated. If the plastic is made from petrochemicals and incinerated, then fossil resources are being converted to carbon dioxide emissions. If the plastic is PLA and it is incinerated, then the carbon is *converted back to* CO_2. If the product does not enter the waste disposal system, then the time horizon becomes important. In the case of permanent products, for a relatively short time frame such as a human lifetime, I suppose the products are permanent. For long time horizons, such as hundreds, thousands, or millions of years, then my assumptions regarding permanent products may change. At these longer time scales, I suspect that even permanent thermoplastics may get converted to carbon dioxide. In our LCI analysis, we have used shorter time frames, on the scale of decades, and made conservative assumptions.

In our LCI, we have tried to be as complete as possible, but all must realize that we are, in fact, using LCI as a tool to guide our developments. Our LCI profile will change depending on what we learn and the resulting new directions we take, and our energy use could even be higher for a period of time.

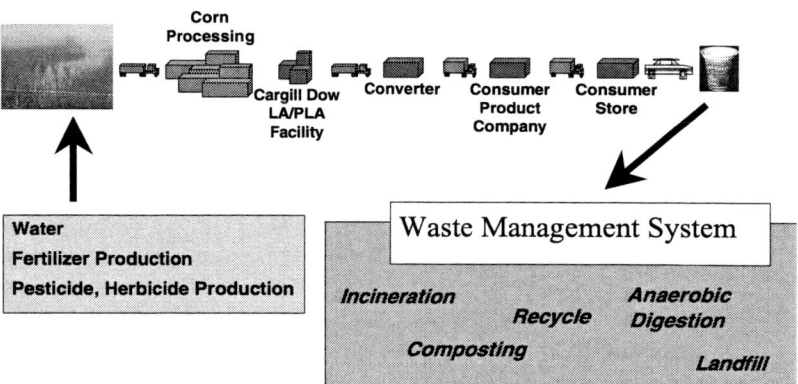

FIGURE 11.10 Cradle-to-grave understanding requires knowledge across all parts of the business system.

However, the general trend over time will be lower energy and lower carbon dioxide emissions. Our long-term goal over the long run is to eliminate petrochemicals from our business systems. We believe we will be rewarded in the marketplace.

Figure 11.11 compares total petroleum (feedstock + process energy) for production of conventional polymer pellets versus projected total petroleum used for production of PLA polymer pellets. The polymers listed are expected to compete with PLA in the marketplace in some specific applications. "PLA year 1" data were generated using assumptions based on the process technology to be deployed in Cargill Dow's production facility currently under construction. "PLA year 5" and "long-term" data assume improvements in process technology and alternative energy sources that further optimize production efficiencies for PLA. CD is already investing in the technology to reach the "year 5" goals.

It is clear in Figure 11.11 that PLA has the advantage in total fossil resource use. Note that although cellophane is made from renewable resources, its fossil resource use is higher than many of the other products. However, assessing fossil resource use only up to polymer pellet production is incomplete. For a complete LCI comparison, converted-product and end fate data must be included. The results of fossil resource use in this type of analysis are shown for PLA versus polystyrene cups and PLA versus polyethylene terephthalate (PET) bottles in Figures 11.12 and 11.13, respectively.

PLA is advantaged in each comparison. It is interesting to note that the fossil resource differs with each of the disposal methods. Neither polystyrene nor PET can be composted or degraded in anaerobic digestion.

Fossil resource use drives CO_2 emission. The more process energy used, the more carbon dioxide is emitted. If incineration is the end fate, then the carbon of the material itself is emitted as carbon dioxide. If landfill is the end fate, then CO_2 emissions on disposal arise primarily from transportation. If composting is the end fate, then most of the carbon is converted to CO_2 (we've assumed 95%), and the rest is humic substance.

The results of the cradle-to-pellets analysis for CO_2 emissions from conventional plastics and PLA are summarized in Figure 11.14. "PLA year 1" data represent the production facility currently under construction by Cargill Dow and projected to be on-line in late 2001. "PLA year 5" and "long-term data"

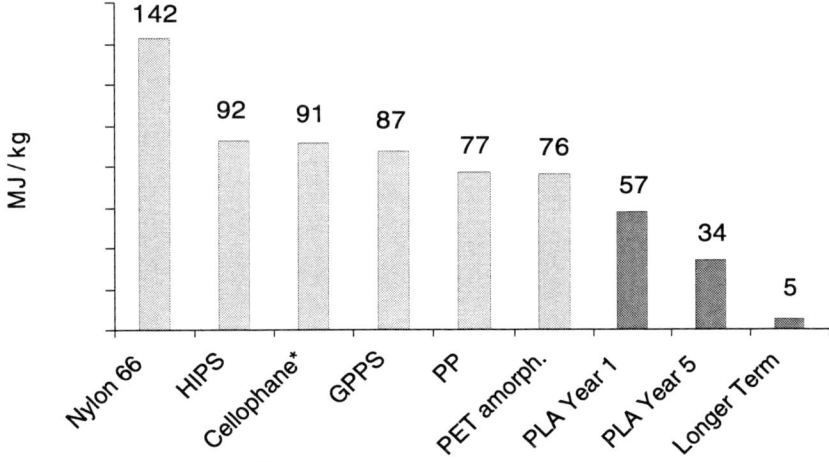

FIGURE 11.11 Total fossil resource use in common plastics measured in energy. All of the LCI results are based on the Boustad model.[3]

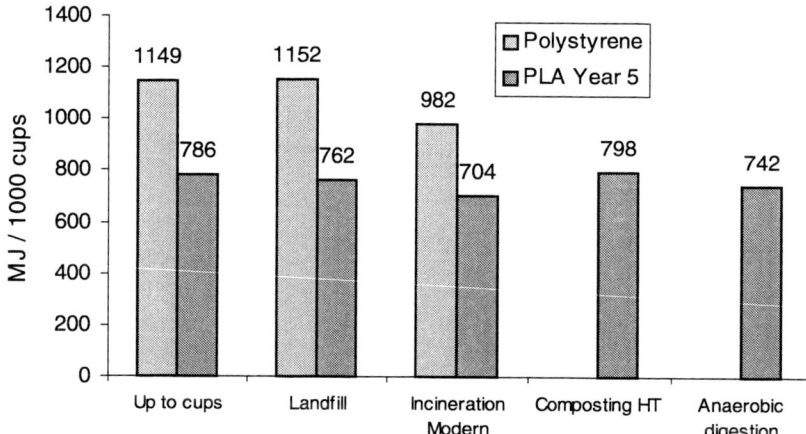

FIGURE 11.12 Fossil resource use in PLA cups compared to polystyrene cups through various waste disposal systems.

assume improvements in process technology and alternative energy sources to further reduce CO_2 contributions by PLA.

Again, using the analysis only through pellets is incomplete. Figure 11.14 shows that at a pellet level PLA can be a carbon dioxide sink! However, when the complete comparison of PLA to polystyrene cups (Figure 11.15) and PLA to PET bottles (Figure 11.16) is done, we find that PLA is still advantaged, with a greater than 50% reduction in CO_2 emissions relative to conventional plastics.

The primary reason that PLA has a favorable carbon dioxide profile is that the fermentation and chemical processing steps are extremely efficient. In our analysis, we charged all of the agricultural

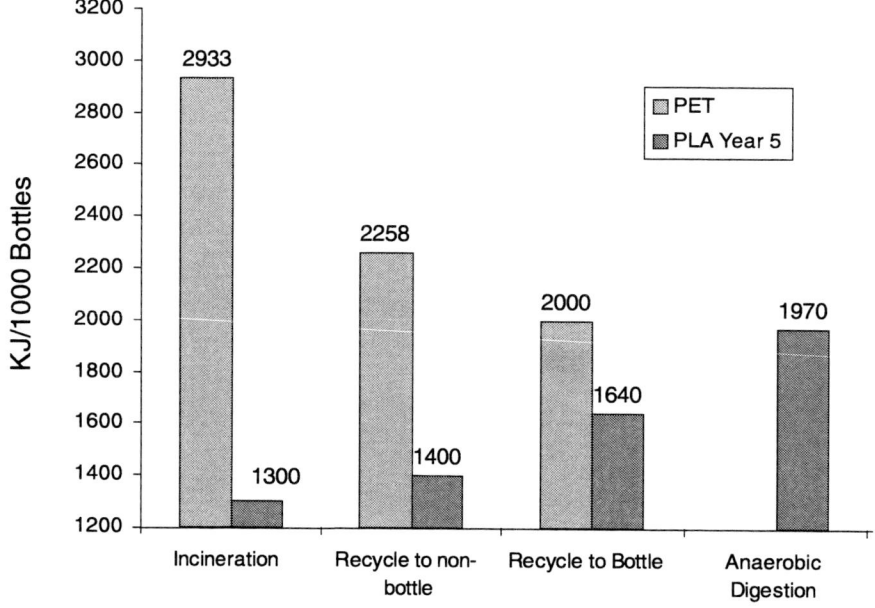

FIGURE 11.13 Fossil resource use in PLA bottles compared to polystyrene bottles in various waste disposal systems.

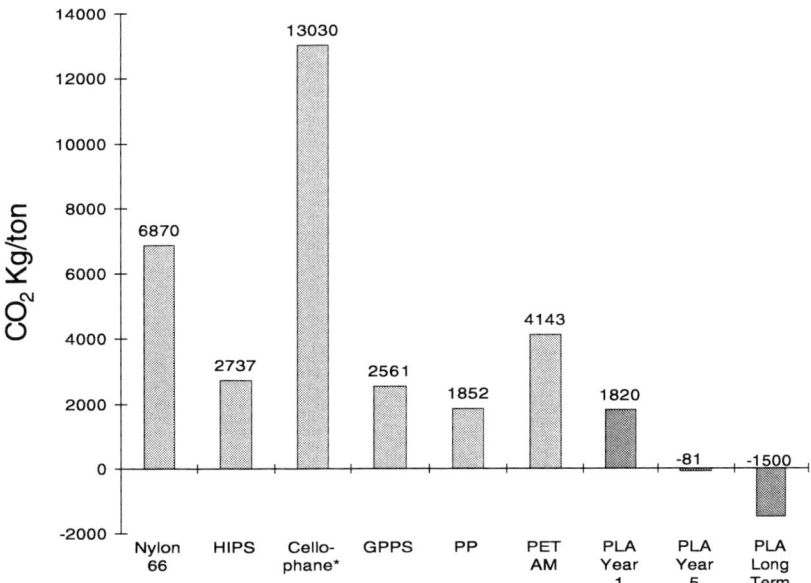

FIGURE 11.14 Net CO_2 emissions of PLA compared to other polymers: cradle to pellets.

inputs and emissions to dextrose. If our PLA yield was lower, not only would the per-cup process energy inside our fence increase, but all of the upstream inputs would increase due to the additional corn and farming needed. The take-home lesson is that yield is particularly important.

High concentration of product in the fermentation is also important for efficient bioprocessing. In the case of lactic acid, the concentration (titer) is typically greater than 100 g/L. This means that less water has to be removed from the product—hence, less energy input to evaporation and lower carbon dioxide emissions.

To my thinking, a general rule would be that a commercially viable industrial product produced from renewable resources using fermentation and biotechnology must have yields greater than about

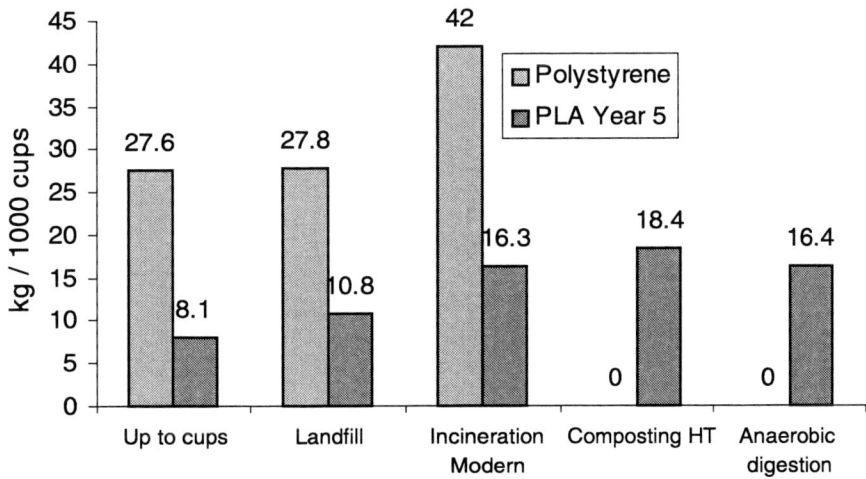

FIGURE 11.15 Carbon dioxide emissions of PLA cups compared to polystyrene cups on a cradle-to-grave basis.

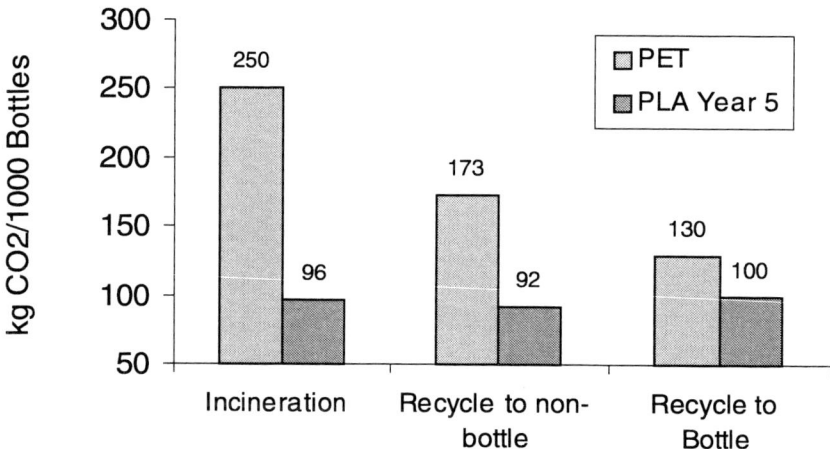

FIGURE 11.16 Carbon dioxide emissions of PLA bottles compared to PET bottles on a cradle-to-grave basis.

85% and a titer of about 90-100 g/L in order to have a carbon dioxide and fossil resource advantage compared to petrochemical-based products.

Composting and Biodegradation of PLA

PLA is unusual in that it is stable under normal-use conditions but degrades quickly in environments of high temperature, high moisture, and high microbial activity. This feature may allow accelerated development of alternative waste disposal systems, such as composting and anaerobic digestion.

The primary mechanism of degradation is hydrolysis, catalyzed by temperature, followed by bacterial attack on the fragmented residues. The environmental degradation of PLA occurs by a two-step process. During the initial phases of degradation, the high-molecular-weight polyester chains hydrolyze to lower-molecular-weight oligomers. This reaction can be accelerated by acids or bases and is affected by both temperature and moisture levels. Embrittlement of the plastic occurs during this step at a point where the number average molecular weight (M_n) decreases to less than about 40,000. At about this M_n, microorganisms in the environment continue the degradation process by converting these lower molecular-weight components to carbon dioxide, water, and humus. The structural integrity of molded PLA articles decreases as the molecular weight drops, and eventually the article disintegrates. A typical degradation curve of PLA under composting conditions is shown in Figure 11.17.

Under typical use and storage conditions, PLA products are stable. In addition, certain additives can be used to retard hydrolysis. Continuing studies in this area will lead to increased PLA stability for more extreme applications.

Research and Technology

PLA is interesting in that in everything we do to lower the cost of our product, we gain an enhancement in our LCI profile. If we lower our energy use, our LCI profile gets even better. If we use alternative feedstock's (biomass) as a raw material, we eliminate the processing energy and chemical use required in wet milling of corn. Our long-term goal is to eliminate fossil fuel resource use, the source of carbon dioxide emissions.

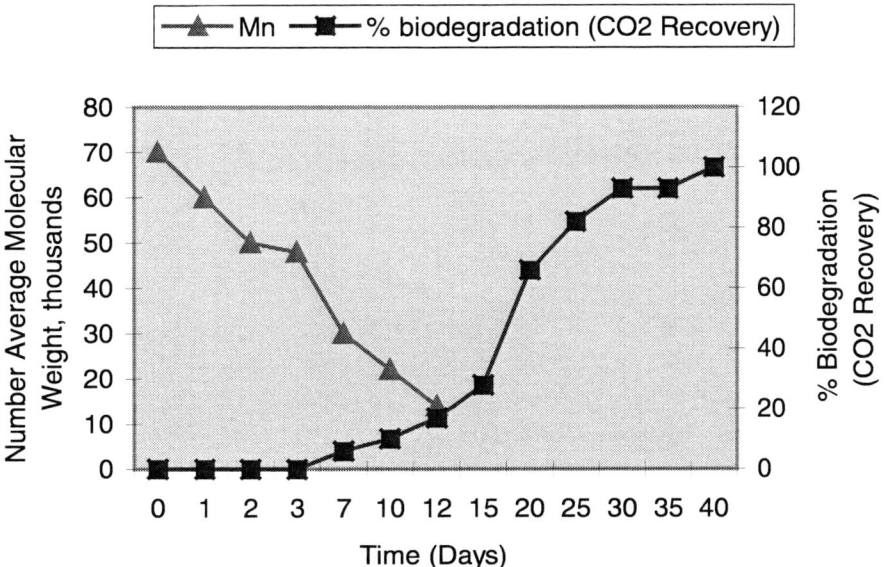

FIGURE 11.17 Biodegradation of PLA in compost at 60° C. NOTE: M_n = number average molecular weight.

On the product side of technology development, we have room to expand our product performance and market potential. The stereoisomers of PLA allow a very wide variety of performance—we have not yet explored this fully. We also will begin to target other molecules made from lactic acid. In our view, lactic acid will become a very inexpensive raw material used in a wide range of products.

We will continue to modify our LCI and use the data to assist our customers to make choices to reduce their environmental impact. We find ourselves forced to develop appropriate agricultural data, because the practices that farmers employ impact our environmental profile.

SUMMARY

PLA uses fewer fossil resources and emits less carbon dioxide in its manufacturing than the petrochemical-based products it replaces. The differences are measurable and significant according to the standard methodology for LCI. PLA fits into any waste management system, including composting or other managed biodegradation systems.

The expectation and perception associated with renewable resources are that the products should be advantaged compared to the petrochemical products they are replacing. Our data suggest that this perception applies—at least for PLA made by our process.

Here are some lessons we have learned with PLA:

1. Research and development of environmentally friendly technology must result in products that have value in the marketplace. The products must perform in the intended application. No one will buy a product that doesn't perform, even if it is environmentally friendly. This ought to be obvious, but it is often overlooked. PLA has an opportunity because it performs well at competitive price. It has a very large market potential that should develop rapidly because it uses fewer fossil resources, emits less carbon dioxide, and is compostable. Furthermore, if we market it properly, PLA will help consumers understand that high-performing products can be made from renewable resources and that they are, in

fact, better for the environment. Companies like ours are changing the paradigms and expectations of consumers, while creating more demand for products with improved environmental performance.

2. CO_2 is a raw material, not a waste product. PLA uses it successfully as a raw material because the fermentation and chemical processing technologies are very efficient and have high yields. What additional technologies can create other useful products from CO_2? Can fermentation technology be developed to fix CO_2?

3. Successful renewable resource-based products will have to combine fermentation and biotechnology with chemistry. Combining fermentation and biotechnology with chemistry can create special advantages and new opportunities because nature is very good at making compounds that are difficult to synthesize by traditional chemistry. Chiral lactic acid is a simple example of this dual requirement. Lactic acid has only limited market potential as lactic acid, but through chemistry we can create significant market opportunities that take advantage of the chirality that the natural system provides.

4. We need scientists who can move effectively across disciplines and who know chemistry. Chemists can be extremely effective in a wide variety of technology areas. Chemists are trained in the first principles of reactions and reaction mechanisms. In my experience, they are the ones who understand and are most effective at determining what the critical issues are and what the targets should be. We have been most successful with people who have a depth of chemistry experience and a breadth across related disciplines such as biochemistry, molecular biology, chemical engineering, and business. We need more people who understand the fundamentals of chemistry. I wonder if these basic skills are being lost because of the emphasis on biotechnology.

5. In order for us to reach our long-term environmental goals, we need advancement in biomass processing and alternative energy sources. The economics are easy to see in biomass processing, but the technologies need to be developed. For successful development, a wide variety of companies and institutions will be required to collaborate. Technologies such as wind power have to be fully developed and deployed. We need more renewable energy sources for electricity and steam. We need alternative fuels for transportation. Then we can really reduce carbon dioxide emissions.

6. PLA typifies technology and product development for thermoplastics and other commodity chemicals made from renewable resources. Uncovering performance attributes and value across these business systems takes large investments in R&D. The market developments are long and complicated. The process technology development cuts across many fields and requires an integration of knowledge. Very few companies can undertake and sustain the effort. Development of more renewable resource-based products requires large investments in R&D and in manufacturing facilities. Government funding is critical for identifying issues and developing basic technologies to resolve these issues. In our case, both the Department of Energy (DOE) and the National Institute of Standards and Technology (NIST) funded basic technology that helped us. However, if Cargill, Incorporated, and the Dow Chemical Company had not funded us, I doubt that we would have been able to sustain the effort. Significant government funding will be required if we want to see more products like PLA developed and commercialized.

ACKNOWLEDGMENTS

Mark Hartmann, Nicole Whitemann, Erwin Vink, James Lunt, Kevin McCarthy, Lang Phommahaxay, Ray Drumright, David E. Henton, Patrick Smith, and Jed Randall, Tom Bremel, NIST, and DOE. NatureWorks PLA is a registered trademark of Cargill Dow LLC. The graphics included in this chapter are used with the permission of Cargill Dow LLC.

REFERENCES

1. Cargill Dow LLC's Web site provides a summary at www.cdpoly.com.
2. There are many references that discuss the material properties of PLA and PLA stereo-polymers. The following list provides good background reading:
   Ninjenhuis, A.J., D.W. Grijpma, and A.J. Pennings, 1991, *Polym. Bull* 26:71. R. Vasanthakumari and A.J. Pennings, 1983, *Polymer* 24:175. B. Kalb and A.J. Pennings, 1980, *Polymer* 21: 607. G.L. Loomis, and J.R. Murdoch, *U.S. Patent 4902515*, 1990. M. Spinu, *U.S. Patent 5317064*, 1994. H. Tsuji and Y. Ikada, 1993, *Macromolecules* 26:6918. Y. Ikada, K. Jamshidi, H. Tsuji, and S.H. Hyon, 1987, *Macromolecules* 20:904. J. J. Kolstad, 1996, *J. Applied Polym. Sci.*62:1079. J. Huang, M.S. Lisowski, J. Runt, E.S. Hall, R.T. Kean, N. Buehler, and J.S. Lin, 1998, *Macromolecules* 31:2593. E.W. Fischer, H.J. Sterzel, and G. Wegner, 1973, *Kolloid-Z.u.Z. Polymere* 251:980. R. von Oepen and W. Michaeli, 1992, *Clin. Mat.* 10:21. G. Kokturk, T. Serhatkulu, A. Kozluca, E. Piskin, and M. Cakmak, 1999, *Annu. Tech. Conf. – Soc. Plast. Eng.* 57(2):2190. R. Sinclair, 1996, *J. Macromol. Sci. Pure Appl. Chem.* A33(5):585. R. Sinclair, *U.S. Patent 5424346*, 1995. X. Chen, K. Schiling, and W. Kelly, *U.S. Patent 5756651*, 1998. S. McCarthy, R. Gross, and W. Ma, *U.S. Patent 5883199*, 1999. M. Spinu, C. Jackson, M. Keating, and K. Gardner, 1996, *J.M.S. – Pure Appl. Chem.* A33(10):1497. D. Verser, K. Schiling, and X. Chen, *U.S. Patent 5633342*, 1997. S. Harper, *U.S. Patent 5032671*, 1991. J.R. Dorgan and J.S. Williams, 1999, *J. of Rheology 43(5)*:1141. P. Gruber, J. Kolstad, D. Witzke, M. Hartmann, and A. Brosch, *U.S. Patent 5594095*, 1997. P. Gruber, J. Kolstad, D. Witzke, M. Hartmann, and A. Brosch, *U.S. Patent 5798435*, 1998. P. Gruber, J. Kolstad, D. Witzke, M. Hartmann, and A. Brosch, *U.S. Patent 5998552*, 1999. C. Ryan, M. Hartmann, J. Nangeroni, 1997, *Polymers, Laminations, and Coatings Conf.* 139.
3. Boustad, I., "Eco-Profiles of Plastics and Related Intermediates, Methodology," The Association of Plastics Manufacturers in Europe, April 1999. Report is available at http://www.apme.org.
4. Kamath, Y.K. S.B. Hornby, H.D. Wiegmann, and M.F. Wilde. 1994. Wicking of spin finishes and related liquids into continuous filament yarns. *Textile Res. J.,* 64(1)33-40.

DISCUSSION

Tobin Marks, Northwestern University: When you go from glucose by fermentation to lactic acid, is that a continuous process or a batch process?

Patrick Gruber: We can do it either way. We haven't described the way we *will* do it to anyone yet. It can be done effectively either way. Parts of the process will probably be done in batches, but some of it depends on how "batch" is defined. Overall, though, it is a continuous process.

The lactic acid must be isolated and purified by removing water and taking out impurities. Then to get to PLA from there involves straight chemistry. As the water is being evaporated from the lactic acid, it starts to do a condensation reaction, because you have both a hydroxyl group and an acid group to give what we call "pre-polymer oligomers." These start to form an equilibrium mixture with the lactide through a back-biting reaction. In contrast, if we wanted to make conventional polyester, we would heat it and apply a vacuum to condense it. If, at this point, we tried to condense lactic acid directly to make high-molecular-weight material, a dimer, lactide, would be clipped off. We cross over the conditions where the lactide is vaporized—high enough temperatures and low enough pressures. We purify the lactide ring in a distillation column. This allows us to control even the optical isomers in the lactates. We then collect the lactides. Lactide can be thought of as an anhydrous lactic equivalence in a ring form. We simply do a ring opening polymerization to form the polymer. It is quite an efficient process.

Tom Baker, Los Alamos National Laboratory: Can you tell us something about the kind of footprint you would need for plants when PLA cup production becomes very large? In other words, if you are going to make a hundred thousand tons of polymer a year using biocatalysis, what kind of footprint are you going to need?

Patrick Gruber: I think making a billion pounds of polymer would require a farm producing corn that is about 10 miles on any side. For us to make 300 million pounds of plastic would require about 400 million pounds of lactic acid. Assuming 120 bushels per acre, this equates to 117,000 acres of corn.

Rosemarie Szostak, Department of the Army: To follow up on that and address the aspect of food security, what percentage of the corn market are you anticipating this to be?

Patrick Gruber: The percentage of corn that we would impact is virtually insignificant. The amount of corn produced in the United States is so massive—7 to 9 billion bushels per year (56 pounds per bushel)—and the amount used for making these polymers is so small that it doesn't even register. A fraction of one corn plant could supply us a billion pounds. It is minuscule in the grand scheme of corn.

Rosemarie Szostak: How do you anticipate this in an international market with regard to agriculturally at-risk areas and deforestation—for example, in developing countries?

Patrick Gruber: Third World countries are trying to figure out what infrastructure they should build. This includes chemical processing systems, bio-processing systems, food processing systems and, of course, waste disposal systems. What can't be done is to take land required for food production, or to further degrade marginal land. This is an issue we need to pay attention to. For example, the amount of material that would feed our PLA plant, would take all the export sugar of Australia. I know they are not a third world country, but that gives you an idea of size. Australia is a good sugar producing country. We would need all their export sugar capacity just to feed our little plant. Our manufacturing plant and raw materials could impact agriculture land and its use. We have to pay attention.

The idea of using energy crops in third world countries as raw materials is hard for me. From an economic point of view it is disadvantaged because you wind up with a low energy concentration in the agricultural product, and therefore, more land. In Third Worlds, you need food production, hence it can only make sense in parts of the world where food is not an issue.

David Keith, Carnegie Mellon University: I want to sound a somewhat cautionary note about the life-cycle analysis. You implied that if you could do a life-cycle analysis and show that some process was producing less carbon, you might get a tax credit for it.

Even with a lot of work, in the absence of bias, it is extraordinarily difficult to be sure that the results of a life-cycle analysis are correct. This is one of the reasons that many folks would argue very strongly that we should not institute a system of tax credits in that way. The accounting challenges are formidable; firms would have an incentive to exaggerate the results, and there is no gold standard to hold them accountable to. I think it makes more sense to tax the relevant environmental insults directly—carbon dioxide emissions or nitrous oxide emissions, or whatever. If you do it that way, you get a very different answer. No longer does a company have to produce some giant statement that the government then has to audit, but the company just has to minimize cost, the way it always did, under appropriate prices.

To give you one example, corn production yields a lot of N_2O and methane. So, it is a big net greenhouse gas emitter. Maybe that effect was included, but there are probably seven others that were not.

Patrick Gruber: To clarify: what I said was that if there were a credit available, we would take advantage of it. The market opportunity is created solely because the product works well, customers are willing to buy it, and it is economical. This fundamental principle is too often lost. We are going to take advantage of tax credits or of other markets who are penalized through taxes.

This is the way it actually will unfold. Your point about the life-cycle studies is well taken. It is extremely problematic to sort through these data. You saw me actually cast the argument through this global system of comparing oil, agricultural use, getting it down to a yogurt cup. If we are going to talk about a thousand yogurt cups, this is actually the difference. Now, I am more confident about the data from inside my fence line, but it gets complicated outside the fence line.

Alan M. Wolsky, Argonne National Laboratory: You remarked that the next target would be to make styrene and related compounds from lignin. I am confused about whether or not that input lignin would otherwise be used as fuel. If so, as is the case in most pulp manufacture, is there a benefit to the climate in forgoing the use of lignin as a renewable fuel, and, instead, turning it into benzene, toluene, and xylenes?

Patrick Gruber: That is a good question and ought to be answered, but I don't know the answer to it. My guess is that there will be a set of chemicals that can be made from lignin that have particular value, and these ought to be the ones that get made. The others ought to be burned if the better value is to burn them.

Richard Wool, University of Delaware: Where, along the chain of reactions, are you expending your highest fossil fuel energy, in terms of conversion processes?

Patrick Gruber: In the fermentation process and, more specifically, dealing with the water that is there. That is why I talked about the criteria of what a good fermentation looks like. You can have either too low titer of product or too low yield.

Richard Wool: Is that mostly from the dextrose itself or is that from the starch? Are you talking about starch to lactic acid going through a glucose breakdown process?

Patrick Gruber: The glucose systems are pretty effective. For fermentation products, the same goes for the lactic acid system. It could be improved by moving up the finished product titer. The energy use is determined by the amount of water that has to be evaporated.

Richard Wool: Is that a very tight system or do you think it is open to biotechnological improvements in the future?

Patrick Gruber: It is open to biotechnology improvements, for sure. When I show how we will progress to the future, there is no doubt in my mind that we can impact all critical parameters.

Richard Wool: Is the bottom line, even right now, that you have a considerable CO_2 advantage over the fossil fuel?

Patrick Gruber: If we compare them that way, with the caveats we talked about, but following the rules, then yes it does. So I would say that we start off in a good place. We can start to establish this category in the marketplace, and we can learn how to do the LCI or whatever the measurement technique is. There is no doubt that as we apply new biological techniques, catalyst programs, and chemistry, we will make it even better.

Richard Wool: How much fossil fuel energy is expended in breaking the starch down into sugars?

Patrick Gruber: I have the numbers for total percent of the PLA, but I don't remember exactly. It is very small, because it is a highly efficient enzymatic process that takes place under mild conditions.

12

Chemicals from Plants

John W. Frost, K.M. Draths, David R. Knop, Mason K. Harrup,
Jessica L. Barker, and Wei Niu
Michigan State University

Virtually all pseudo commodity and commodity chemicals as well as most fine chemicals are synthesized from petroleum feedstocks. It has been estimated that 98% of all chemicals produced in the United States in excess of 2×10^7 kg are synthesized from petroleum and natural gas.[1] By contrast, chemicals isolated as natural products from plants or produced by microbes from carbohydrate feedstocks are typically restricted to ultrafine chemicals and a relatively few fine chemicals. The goal of our research effort is to ascertain how the widest possible spectrum of commodity, pseudo commodity, fine, and ultrafine chemicals can be synthesized from polyol starting materials such as d-glucose, d-xylose, l-arabinose, and glycerol. These starting materials, in turn, are derived from renewable feedstocks derived from plants such as starch, hemicellulose, cellulose, and oils. A key feature of these conversions is the use of recombinant microbes as synthetic catalysts.

The implications of switching chemical manufacture from its current reliance on nonrenewable fossil fuel feedstocks to utilization of renewable, plant-derived feedstocks is considerable. The U.S. chemical industry currently uses 5 quads (1 quad = 1×10^{15} British thermal units) of carbon for manufacture of organic chemicals.[2] If these 5 quads of carbon were derived from renewable feedstocks, a net consumption of CO_2 would be realized, given that the polyols used as starting materials are biosynthesized by plants from CO_2 and are essentially immobilized forms of CO_2. Since international treaties eventually require reductions in CO_2 emissions in the United States, large-scale consumption of CO_2 to make chemicals may be used as a credit to offset CO_2 emissions resulting from combustion of fossil fuels to generate electricity and combustion of petroleum-based transportation fuels.

Synthesis of chemicals from CO_2-derived feedstocks requires that an understanding be developed of the factors that influence the yield and titer of chemicals that are microbially synthesized from polyol starting materials. Effective interfacing of microbial catalysis with chemical catalysis is also critical to being able to synthesize the widest possible spectrum of chemicals from renewable polyols. Finally, the diversity of molecules that can be microbially synthesized from polyols is ultimately dependent on the diversity of microbes, biosynthetic pathways, and genes encoding biosynthetic enzymes that are available to synthetic chemistry.

YIELD AND TITER CONSIDERATIONS

GS4104 is a neuraminidase inhibitor that is effective as an anti-influenza drug.[3] Currently marketed by Roche as Tamiflu, the key problem with the synthesis of this drug was the availability of the hydroaromatic starting material. Shikimic acid is an obvious candidate as a starting material; it is a cyclohexene carboxylic acid possessing the appropriate asymmetric centers. However, only limited quantities of shikimic acid were available because its isolation from the seeds of *Illicium* plants is expensive.[4] The unfavorable price and limited availability of shikimic acid have been major impediments to its use by chemists as a chiral synthon. Quinic acid, which is isolated from *Cinchona* bark,[4] is less expensive and more available. This hydroaromatic has been widely employed by chemists as a chiral synthon,[5] although considerably more steps are required to convert quinic acid into GS4104 (Figure 12.1) relative to the use of shikimic acid as starting material.[3] The scarcity and expense of shikimic acid are similar to those of numerous natural products that are important pharmaceutical building blocks because they too have to be isolated from plants whose cultivation does not benefit from large-scale monoculture.

Our task was to design and construct a microbe that would synthesize shikimic acid.[6] *Escherichia coli* was selected as the starting point for construction of a shikimate-synthesizing biocatalyst. Shikimic acid is an intermediate in the common pathway of aromatic amino acid biosynthesis (Figure 12.2), which begins with the condensation of phosphoenolpyruvic acid with d-erythrose 4-phosphate catalyzed by 3-deoxy-d-*arabino*-heptulosonic acid 7-phosphate (DAHP, Figure 12.2) synthase. DAHP is then cyclized to 3-dehydroquinic acid (DHA) by *aroB*-encoded 3-dehydroquinate synthase. Dehydration of 3-dehydroquinic acid catalyzed by *aroD*-encoded 3-dehydroquinate dehydratase is followed by the reduction of the resulting 3-dehydroshikimic acid (DHS) to the desired shikimic acid by *aroE*-encoded shikimate dehydrogenase. To prevent the conversion of shikimic acid into shikimate 3-phosphate, both isozymes of shikimate kinase were inactivated by transduction using P1 phage of *aroL*478::Tn*10* and *aroK17*::Cm^R. The next task was to increase the flow of carbon down the common pathway. This was accomplished by plasmid localization of a mutated isozyme of DAHP synthase encoded by *aroF*^{FBR} that is insensitive to feedback inhibition by aromatic amino acids. By blocking the conversion of shikimic acid into shikimate 3-phosphate, de novo biosynthesis of aromatic amino acid is blocked. l-Phenylalanine, l-tyrosine, l-tryptophan, and aromatic vitamins must be added to cultures to enable the microbial construct to grow, but they inhibit DAHP synthase activity and thus the flow of carbon into the common pathway. Employment of *aroF*^{FBR} and the encoded DAHP synthase that is insensitive to feedback inhibition is thus critical to directing increased carbon flow into aromatic amino acid biosynthesis.[6]

Individual common pathway enzymes become impediments to increased carbon flow due to their inability to convert substrate into product at a rate sufficiently rapid to avoid cytoplasmic accumulation and subsequent export of the substrate into the culture medium.[7] 3-Dehydroquinate synthase and shikimate dehydrogenase are both impediments to increased carbon flow that result, respectively, in accumu-

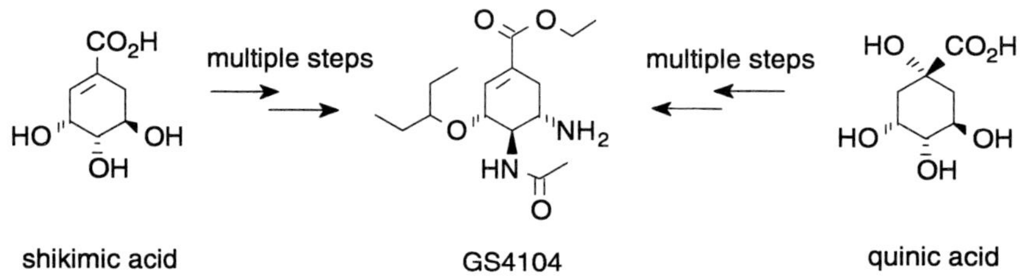

FIGURE 12.1 Starting materials for the synthesis of the neuraminidase inhibtor GS4104.

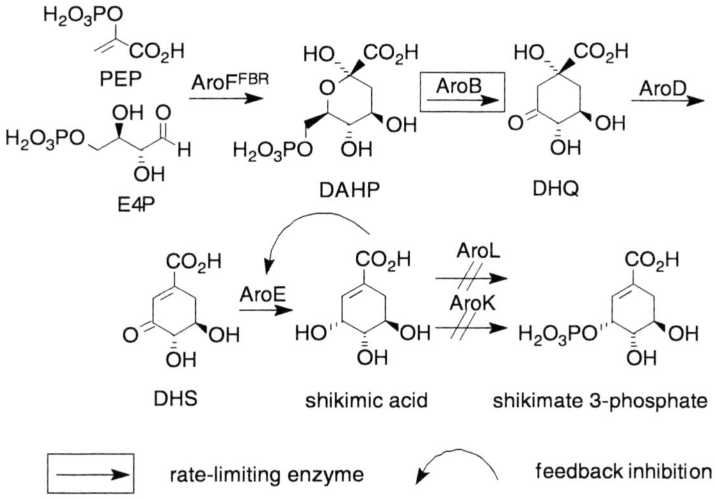

FIGURE 12.2 Shikimic acid biosynthesis.

lation of 3-deoxy-d-*arabino*-heptulosonic acid (DAH) and 3-dehydroshikimic acid (DHS in Figure 12.2).[7] These accumulating metabolites reduce the yield and titer of the desired shikimic acid and introduce contaminants that require additional purification of product shikimic acid. Accumulation of DAH is eliminated by site-specific insertion of a second *aroB* locus encoding DAHP synthase into the genomic *serA* locus.[7b] The resulting increase in the specific activity of 3-dehydroquinate synthase eliminates DAH accumulation. Accumulation of 3-dehydroshikimic acid results from feedback inhibition of *aroE*-encoded shikimate dehydrogenase by shikimic acid.[7a] A portion of this inhibition can be circumvented by the increase in shikimate dehydrogenase-specific activity attendant with plasmid localization of *aroE*. Site-specific insertion of *aroB* into the genomic *serA* locus also provides for plasmid maintenance. The only way the final shikimate-synthesizing construct can grow in minimal salts medium lacking l-serine supplementation is to retain the plasmid encoding the *serA* locus along with the *aroE* and *aroF*^{FBR} loci.

The *E. coli* SP1.1/pKD12.112 construct incorporating all of the aforementioned genetic elements was cultivated under batch-fed fermentor conditions at 33°C, pH 7, and with dissolved O_2 maintained at 10% of air saturation. Under these conditions, *E. coli* SP1.1/pKD12.112 synthesized 20.2 g/L of shikimic acid in 14% yield.[6] Although 1.9 g/L of quinic acid was produced, this amount was sufficiently low that it could easily be removed by recrystallization. To put the yield of shikimic acid achieved thus far into perspective, it is important to note that besides the in vivo activity of DAHP synthase, carbon flow directed into the common pathway is also dictated by the availability of d-erythrose 4-phosphate and phosphoenolpyruvic acid. By improving the availability of these substrates, the theoretical maximum yield for *E. coli*-catalyzed synthesis of shikimic acid from glucose is 86%.[8] Achieving such yields would likely move shikimic acid from the realm of a scarce natural product used as a chiral synthon to the realm of an inexpensive building block chemical possessed of disposable chirality and an attractive assemblage of carbon atoms.

INTERFACING AND BIOCATALYSIS

Although shikimic acid and quinic acid are certainly not large-volume chemicals, these hydroaromatics provide an intriguing entry point into large-volume fine chemicals and pseudocommodity chemicals. Early syntheses of a number of chemicals employed hydroaromatic starting materials. For example, Woskresensky reported that hydroquinone was produced upon "dry distillation" of quinic acid in 1838.[9] The "quin" root of hydroquinone reflects this genesis. Eykmann reported in 1891 that refluxing concentrated HCl solutions of shikimic acid resulted in the formation of p-hydroxybenzoic acid.[10] With microbe-catalyzed conversions of glucose into shikimic acid and quinic acid, the opportunity presents itself to synthesize hydroquinone and p-hydroxybenzoic acid from glucose.

p-Hydroxybenzoic acid ($4/kg) is a fine chemical used in the manufacture of dyes, pharmaceuticals, and pesticides. Methyl, ethyl, propyl, and butyl p-hydroxybenzoate (known as parabens) are antimicrobial agents found in food, pharmaceutical, and cosmetic applications.[11] Estimates of the market volume for p-hydroxybenzoic acid are difficult to find, although use of parabens is estimated at 1.4×10^6 kg/per year.[12] An emerging market for p-hydroxybenzoic acid is as a component of liquid crystalline polymers such as Xydar and Vectra.[13] p-Hydroxybenzoic acid is manufactured by Kolbe-Schmitt reaction (Figure 12.3, top) of dried potassium phenoxide with 20 atmosphere (atm) dry carbon dioxide at 180-250°C.[14,15] Product potassium p-hydroxybenzoate is converted to its free acid upon addition of mineral acid. The Kolbe-Schmitt reaction is more typically associated with manufacture of salicylic acid (o-hydroxybenzoic acid), which produces p-hydroxybenzoic acid as a by-product. However, thermal rearrangement of the potassium salt of salicylic acid above 150°C leads to predominant formation of p-hydroxybenzoic acid.

A variety of different acids were examined as catalysts for the dehydration of shikimic acid to p-hydroxybenzoic acid. The best yields for conversion of shikimic acid into p-hydroxybenzoic acid were obtained using trifluoromethanesulfonic (CF_3SO_3H) as the acid catalyst and acetic acid (CH_3CO_2H) as the reaction solvent (Figure 12.3, bottom). With this dehydration, a route for synthesis of p-hydroxybenzoic acid from glucose is established. The utility of employing both biocatalysis and chemical catalysis becomes apparent upon consideration of an entirely microbe-catalyzed conversion of glucose into p-hydroxybenzoic acid. Such a biocatalytic route has the advantage that p-hydroxybenzoic acid is synthesized directly. However, one important disadvantage of this strategy is that p-hydroxybenzoic acid is toxic to microbes. Although the toxicity of p-hydroxybenzoic acid toward microbes is signifi-

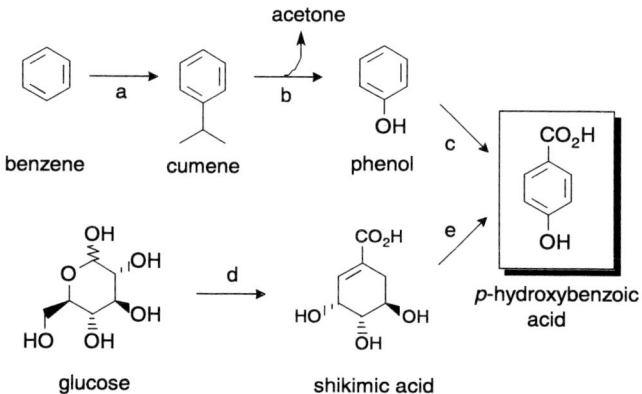

FIGURE 12.3 Synthetic routes to p-hydroxybenzoic acid: top: chemical synthesis; bottom: combined microbial and chemical synthesis.

cantly lower than the microbial toxicity of parabens, it is sufficient to impose a limit on the titer and yield of p-hydroxybenzoic acid synthesized by a microbe directly from glucose. Shikimic acid, in contrast to p-hydroxybenzoic acid, is not toxic to microbes.

Despite the promise of synthesizing p-hydroxybenzoic acid by chemical dehydration of microbially synthesized shikimic acid, there are a number of significant hurdles that have to be surmounted. The yield of p-hydroxybenzoic acid obtained from shikimic acid using CF_3SO_3H as the dehydration catalyst was 75%. In all of the acid-catalyzed dehydrations of shikimic acid, by-products were formed that proved to be difficult to remove without significant loss of product p-hydroxybenzoic acid. Dehydration of shikimic acid required the use of acetic acid as the reaction solvent to avoid the high concentrations of mineral acid (8-12 M) that would be required to catalyze dehydration of shikimic acid in aqueous solution. Acid-catalyzed dehydrations also had to be carried out with purified shikimic acid. Attempts to convert shikimic acid in clarified fermentor broth failed to produce significant yields of p-hydroxybenzoic acid, thus necessitating that shikimic acid be isolated from the fermentor broth prior to its acid-catalyzed dehydration. Additional product loss was encountered during purification of p-hydroxybenzoic acid from acid-catalyzed dehydration of shikimic acid. Combining the yield for microbe-catalyzed conversion of glucose into shikimic acid (14%), the percent recovery of shikimic acid from fermentor broth (86%), and the yield for acid-catalyzed dehydration of shikimic acid after product purification (50%), the overall yield for converting glucose into p-hydroxybenzoic acid is 6%. For comparison, the typical range of yields our group has achieved for direct microbe-catalyzed conversion of glucose into p-hydroxybenzoic acid is 4-13%.

In any multistep synthesis of a chemical product, even relatively small incremental losses in yield in individual reactions, isolations, or purifications have a cumulatively large impact on lowering overall yield. The loss of shikimic acid during isolation from fermentation broth, the crude yield for acid-catalyzed dehydration of shikimic acid, and the loss of p-hydroxybenzoic acid during product purification reduce the overall yield of p-hydroxybenzoic acid synthesized from glucose. Indeed, the overall yield for synthesis of p-hydroxybenzoic acid via acid-catalyzed dehydration of shikimic acid is reduced to the point where this route does not have an advantage in yield relative to direct, microbe-catalyzed conversion of glucose into p-hydroxybenzoic acid. However, significant improvements in the yield of microbe-catalyzed synthesis of shikimic acid can likely be achieved. Homogeneous or heterogeneous, shape-selective acid catalysts that are compatible with use in water at elevated temperatures also need to be elaborated. The availability of such catalysts could improve the yield for conversion of shikimic acid into p-hydroxybenzoic acid, reduce the generation of difficult-to-purify by-products, and avoid the need to use CH_3CO_2H as the reaction solvent.

Hydroquinone is a pseudocommodity chemical ($5/kg) produced globally at volumes of $4.5-5.0 \times 10^7$ kg per year.[16] The major use of hydroquinone is as a photographic developer. Hydroquinone is also employed as a precursor to antioxidants used in rubber and food applications as well as an intermediate in dye manufacture. Oxidation of aniline is the oldest process (Figure 12.4, top) for hydroquinone production and accounts for a relatively modest percentage of global hydroquinone synthesis (approximately 4×10^6 kg/per year).[16] Aniline is initially oxidized by manganese dioxide (MnO_2) in aqueous sulfuric acid (H_2SO_4). Benzoquinone is then reduced by Fe^0 or hydrogenated to afford product hydroquinone. This manufacturing technique generates large quantities of manganese sulfate ($MnSO_4$), ammonium sulfate (($NH_4)_2SO_4$), and iron oxide salts.[16,17]

Hydroperoxidative synthesis[16,17] (Figure 12.4, middle) accounts for approximately 2.5×10^7 kg of hydroquinone production per year. p-Diisopropylbenzene is synthesized by zeolite-catalyzed Friedel-Crafts reaction of benzene or cumene with propylene or isopropanol. Air oxidation of p-diisopropylbenzene proceeds at 90-100°C in an aqueous NaOH solution containing organic bases along with cobalt

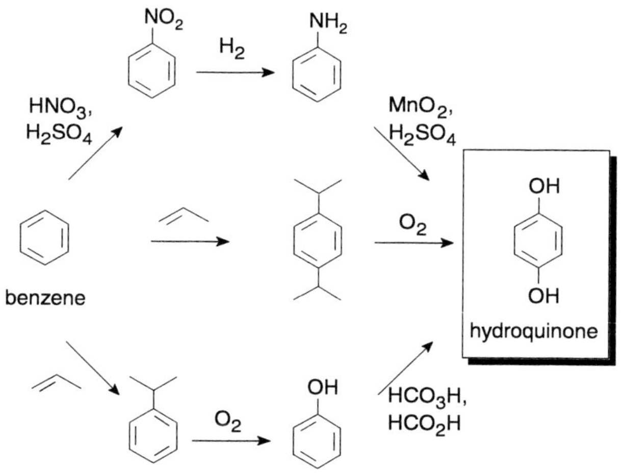

FIGURE 12.4 Hydroquinone synthesis.

or copper salts. Hydroperoxycarbinol and dicarbinol are produced along with the dihydroperoxide during air oxidation. Treatment with acid and H_2O_2 converts the hydroperoxycarbinol and dicarbinol to the dihydroperoxide, which is cleaved to form acetone and hydroquinone. During the acidic cleavage, explosive organic peroxides can form that present a safety hazard.[16]

Reaction of phenol with hydrogen peroxide (H_2O_2) in the presence of strong acids leads to a mixture of hydroquinone and catechol (Figure 12.4, bottom).[16,17,18] This hydroxylation process accounts for approximately 1.4×10^7 kg of hydroquinone production per year. The ratio of hydroquinone to catechol is controlled to a finite extent by the acidity of the catalyst. ZSM zeolites, an aluminosilicate zeolite with high silica and low aluminum content, such as TS-1 are also used as the acid catalyst.[16]

Observations made during synthesis of shikimic acid provided an avenue for elaborating a microbial synthesis of quinic acid (Figure 12.5) that would be appropriate for conversion of glucose into hydroquinone. Along these lines, *E. coli* QP1.1/pKD12.112 was constructed.[6] Host strain QP1.1 was derived by homologous recombination of *aroB* flanked by *serA* sequences into the genomic *serA* locus of *E. coli* AB2848. As with *E. coli* SP1.1, this insertion increases 3-dehydroquinate synthase activity and, by disruption of the *serA* locus, provides a method for maintenance of plasmids containing *serA* inserts. *E. coli* QP1.1 carries a mutation in the *aroD* locus, which renders 3-dehydroquinate dehydratase catalytically inactive. Carbon flow directed into the common pathway of aromatic amino acid biosynthesis can thus not proceed beyond 3-dehydroquinic acid (Figure 12.5). Transformation of *E. coli* QP1.1 with the same plasmid used to construct *E. coli* SP1.1/pKD12.112 leads to *E. coli* QP1.1/pKD12.112. Instead of catalyzing the reduction of 3-dehydroshikimic acid to shikimic acid as it does in *E. coli* SP1.1/pKD12.112, *aroE*-encoded shikimate dehydrogenase catalyzes (Figure 12.5) the reduction of 3-dehydroquinic acid to quinic acid in *E. coli* QP1.1/pKD12.112. Cultivating *E. coli* SP1.1/pKD12.112 under glucose-limited, batch-fed fermentor conditions at 33°C, pH 7, and with dissolved O_2 maintained at 10% of air saturation led, in 60 hours, to the synthesis of 60 g/L of quinic acid in 23% yield.[6] The only by-product that accumulates along with product quinic acid is a relatively small concentration (2.6 g/L) of 3-dehdyroquinic acid.

Although it is theoretically possible to convert glucose into hydroquinone, the fundamental problem with such a route is the toxicity of hydroquinone toward microbes, which is significantly greater than that of *p*-hydroxybenzoic acid. Conversion of glucose into quinic acid and subsequent chemical oxidation of quinic acid to hydroquinone allow hydroquinone's potent toxicity toward microbes to be circum-

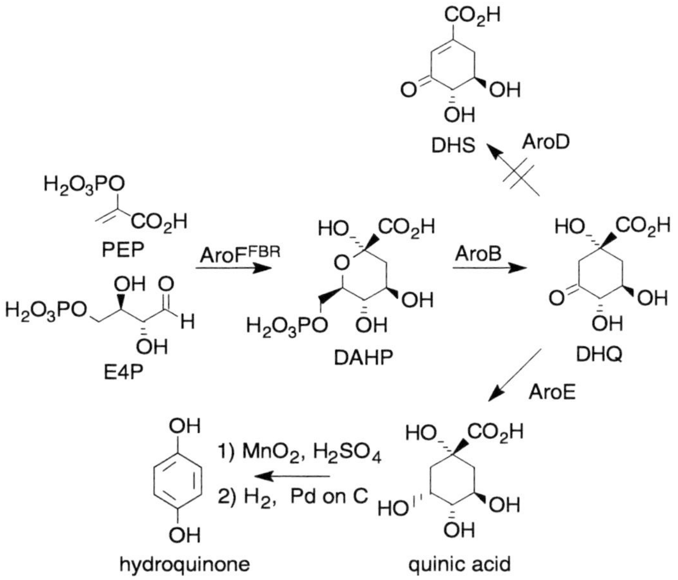

FIGURE 12.5 Combined microbial and chemical synthesis of hydroquinone.

vented. Quinic acid in crude fermentation broth, from which cells have been removed, can be oxidized to benzoquinone in 40% yield after acidification of the broth with H_2SO_4, addition of technical-grade MnO_2, and heating for 1 hour at 100°C.[19] When quinic acid is purified prior to oxidation, 70% conversions of quinic acid to benzoquinone can be achieved.[19] Reduction of benzoquinone affords hydroquinone. Alternatively, aqueous solutions of quinic acid can be converted directly to hydroquinine in 10% yield when heated at 100°C for 18 hours with technical-grade MnO_2 without adding acid.[19]

As with shikimic acid, considerable improvements in the microbe-catalyzed synthesis of quinic acid from glucose need to be and likely can be achieved. The subsequent oxidation of quinic acid to hydroquinone is another area of research in which fundamental advances in catalysis have to be achieved. The current oxidation of quinic acid to benzoquinone using MnO_2 is a stoichiometric oxidation. This runs counter to the current trend toward development of metal-catalyzed oxidations that are catalytic in the amounts of metal required. Although a variety of cooxidants can be employed, elaboration of a metal-catalyzed oxidation in which O_2 is the cooxidant would be particularly desirable. Sufficiently mild conditions for oxidation of quinic acid to hydroquinone may be identified whereby overoxidation to benzoquinone is avoided and hydroquinone is obtained directly from quinic acid. Metal-catalyzed oxidations have to be compatible with use of water as the reaction solvent. Preferably, elaborated catalysts should be sufficiently robust to mediate the oxidation of quinic acid in clarified, crude fermentor broth.

CREATING NEW BIOSYNTHETIC PATHWAYS

Adipic acid is a commodity used primarily in the synthesis of nylon-6,6. Annual global production of adipic acid is approximately 2×10^9 kg.[20] Although a number of new, creative routes have been elaborated for synthesis of adipic acid,[21] most adipic acid manufacture (Figure 12.6) begins with hydrogenation of benzene to cyclohexane.[20] Oxidation of cyclohexane affords a mixture of cyclohexanol and cyclohexanone. This mixture is then further oxidized to afford adipic acid. The nitrous oxide (N_2O),

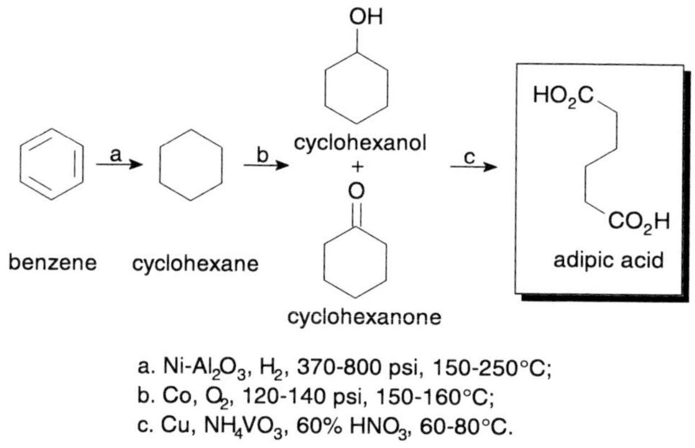
OH
cyclohexanol
+
O
HO2C
CO2H
adipic acid
benzene
cyclohexane
cyclohexanone
a. Ni-Al2O3, H2, 370-800 psi, 150-250°C;
b. Co, O2, 120-140 psi, 150-160°C;
c. Cu, NH4VO3, 60% HNO3, 60-80°C.

FIGURE 12.6 Synthesis of adipic acid.

which is formed as a by-product in the nitric acid oxidation of cyclohexanone and cyclohexanol, is a greenhouse gas and an ozone depleter.[22] Approximately 10% of the annual increase in N_2O levels may arise from this single organic reaction.[22]

As with the synthesis of *p*-hydroxybenzoic acid and hydroquinone from glucose, synthesis of adipic acid from glucose employed both microbial catalysis and chemical catalysis. A recombinant microbe was constructed to synthesize *cis,cis*-muconic acid from glucose (Figure 12.7).[23] Catalytic, chemical hydrogenation of the *cis,cis*-muconic acid then afforded adipic acid (Figure 12.7).[23] Directing increased carbon flow down the common pathway of aromatic amino acid biosynthesis employed the same strategies as previously described for microbe-catalyzed synthesis of shikimic acid. Microbe-catalyzed synthesis of *cis,cis*-muconic acid differed in that 3-dehydroshikimic acid (DHS, Figure 12.6) accumulated instead of shikimic acid. DHS accumulated due to a mutation present in the *aroE* locus that renders shikimate dehydrogenase catalytically inactive in the host *E. coli* strain used for synthesis of *cis-cis*-muconic acid.

3-Dehydroshikimate dehydratase, which is encoded by the *aroZ* locus isolated from *Klebsiella pneumoniae*, catalyzed the dehydration of DHS to form protocatechuic acid (PCA, Figure 12.7). PCA is typically a catabolic intermediate in the *p*-hyroxybenzoate branch of the β-ketoadipate pathway (Figure 12.7). Protocatechuate decarboxylase, which is encoded by the *aroY* locus also isolated from K. *pneumoniae*, then catalyzes the decarboxylation of PCA into catechol (Figure 12.7). Catechol, which is typically encountered in nature as a catabolic intermediate, is distinguished from PCA by virtue of its intermediacy in the separate benzoate branch of the β-ketoadipate pathway (Figure 12.7). *AroY* is a member of a larger family of enzymes that catalyze decarboxylations of benzoic acids possessing electron-donating substituents. Although these enzymes are stable in intact cells, decarboxylase activity is quickly lost in cell-free extracts obtained from cell lysis.

Use of *K. pneumoniae* as the source for isolating *aroZ* and *aroY* is quite advantageous due to the close evolutionary relationship of this microbe to *E. coli*. This relationship means similar codon use for the two microbes and allows both *aroY* and *aroZ* to be expressed in *E. coli* from their *K. pneumoniae* promoters. The last step in the microbe-catalyzed portion of adipic acid synthesis is the catechol dioxygenase-catalyzed conversion of catechol into *cis,cis*-muconic acid, which accumulates in the culture supernatant. Catechol dioxygenase is encoded by the *catA* locus isolated from *Acinetobacter calcoaceticus*. Conversion of glucose into *cis,cis*-muconic acid thus employs genetic elements of three different microbial species.

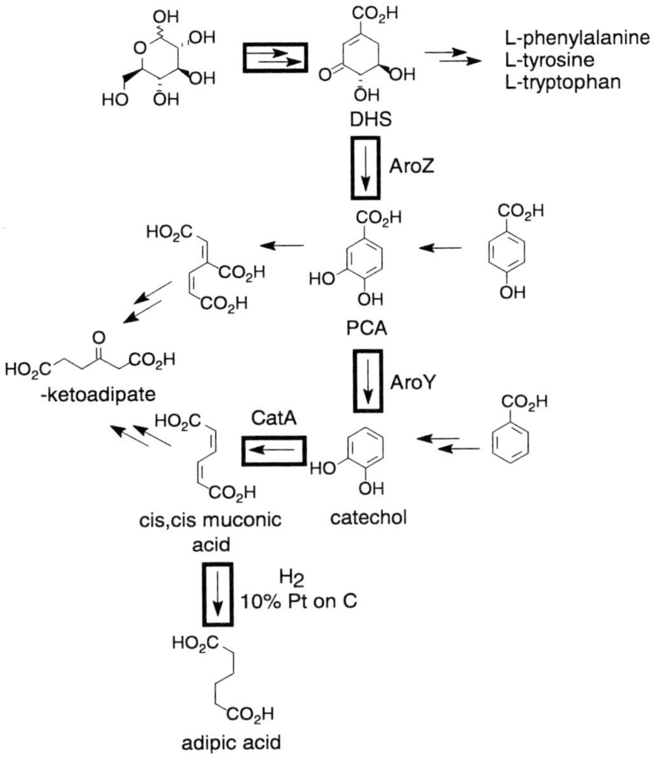

FIGURE 12.7 Combined microbial and chemical synthesis of adipic acid.

In our original work, an *E. coli* host containing three plasmids was used to catalyze the conversion of glucose into *cis,cis*-muconic acid under shake flask conditions.[23] Our most recent construct uses a single plasmid and catalyzes the conversion of glucose into *cis,cis*-muconic acid under controlled, fed-batch fermentor conditions similar to those employed for microbe-catalyzed synthesis of shikimic acid and quinic acid from glucose. After removal of the biocatalyst, the clarified fermentation broth is treated with activated carbon to remove unidentified molecules in the broth that poison platinum catalysts. The *cis,cis*-muconic acid can then be hydrogenated over a platinum-on-carbon catalyst to afford adipic acid in nearly quantitative yield.

For the synthesis of shikimic acid, we elaborated the naturally occurring biosynthesis of a hydroaromatic. The elaborated route to adipic acid, by contrast, has no equivalent in nature. Collecting genes from various microbes and assembling these genes in a single microbial host has created a fundamentally new biosynthetic pathway. Because the starting material possesses the high oxygen content of glucose, the last step of the synthesis is a reduction of *cis,cis*-muconic acid to adipic acid. No by-products are formed during this reduction and no N_2O is generated. This contrasts sharply with use of a deoxygenated starting material such as benzene for adipic acid synthesis. Introducing the requisite oxygen atoms in route to adipic acid requires extensive oxidation. Formation of N_2O is a strategic by-product of such a synthesis.

CONCLUSION

When ongoing genome sequencing and allied functional genomic efforts are combined with the interfacing of chemical catalysis with biocatalysis, a variety of connections can likely be made between carbohydrate starting material and chemical product. Our research is only at the beginning of efforts to delineate the basic connections between plant-derived starting materials and chemical product. This is essentially an exercise in target-oriented chemical synthesis. In progressing from shikimic acid to *p*-hydroxybenzoic acid and then to hydroquinone, we have moved from an ultrafine chemical to a fine chemical and then to a pseudocommodity chemical. Each of these steps constitutes a sizable increase in the volume of the chemical that is manufactured. However, it is not until major commodity chemicals such as adipic acid are synthesized from plant-derived feedstocks that significant progress can be made in transforming the 5 quads of carbon used as starting materials[2] in chemical synthesis in the United States into a form of CO_2 sequestration.

It is important to emphasize that conversion of plant-derived starting materials into chemical products must ultimately lead to industrial-scale processes that yield chemical products at a manufacturing cost that is competitive with their current production from petroleum-derived carbon. Ultrafine chemicals and smaller-volume fine chemicals provide an invaluable proving ground for such activities. Because of the inherent value-added nature of these chemicals, the fundamentals of yield and titer considerations can be elaborated and then scaled up to syntheses commercially practiced by the pharmaceutical and flavor and fragrance industries. The lessons learned will be critical to the large-scale, microbe-catalyzed conversions needed for cost-effective manufacture of pseudocommodity and commodity chemicals from plant-derived feedstocks.

Chemists are uniquely trained to identify and elaborate the connections between plant-derived starting materials and chemical products. At the same time, it is essential that chemists view construction of microbial catalysts as an activity every bit as central to chemical synthesis as development of inorganic and organometallic catalysts. At one time, synthetic chemists were particularly interested in the use of glucose as a starting material in synthetic chemistry. Employment of chiral synthons as starting materials is still an important activity. However, the bioproducts industry is producing an expanding variety of starting materials such as l-lysine and lactic acid at commodity chemical prices and volumes. Transforming this new generation of starting materials into existing and new chemical products requires the development of new syntheses and new synthetic methodologies compatible with use of water and fermentation broths as the reaction solvent. Target-oriented synthetic chemistry and synthetic methodology development thus become equal players with molecular biology and biochemical engineering in the synthesis of chemical products from plants.

REFERENCES

1. Szmant, H. H. P. 4. *Organic Building Blocks of the Chemical Industry*. Wiley: New York, 1989.
2. Bozell, J. J., Landucci, R. *Alternate Feedstocks Program Technical and Economic Assessment*; U.S. Department of Energy, Office of Industrial Technologies: Washington, DC, 1993.
3. (a) Kim, C. U., Lew, W., Williams, M. A., Liu, H., Zhang, L., Swaminathan, S., Bischofberger, N., Chen, M. S., Mendel, D. B., Tai, C. Y., Laver, W. G., and Stevens, R. C. *J. Am. Chem. Soc. 119*, 681, 1997. (b) Rohloff, J. C., Kent, K. M., Postich, M. J., Becker, M. W., Chapman, H. H., Kelly, D. E., Lew, W., Louie, M. S., McGee, L. R., Prisbe, E. J., Schultze, L. M., Yu, R. H., and Zhang, L. *J. Org. Chem. 63*, 4545, 1998.
4. Haslam, E., *Shikimic Acid: Metabolism and Metabolites*. Wiley & Sons: New York, 1993.
5. Barco, A., Benetti, S., De Risi, C., Marchetti, P., Pollini, G. P., Zanirato, V. *Tetrahedron Asymmetry 8*, 3315, 1997.
6. Draths, K. M., Knop, and D. R., Frost, J. W. *J. Am. Chem. Soc. 121*, 1603, 1999.
7. (a) Dell, K. A., and Frost, J. W. *J. Am. Chem. Soc. 115*, 11581, 1993. (b) Snell, K. D., Draths, K. M., Frost, J. W. *J. Am. Chem. Soc. 118*, 5605, 1996.
8. Draths, K. M., and Frost, J. W. *J. Am. Chem. Soc. 117*, 2395, 1995.
9. Woskresensky, A. *Justus Liebigs Ann. Chem. 27*, 257, 1838.
10. Eykmann, J. F. *Ber. Dtch. Chem. Ges. 24*, 1278, 1891.

11. (a) Block, S. S. P. 251. In *Kirk-Othmer Encyclopedia of Chemical Technology 4th Edition*, Volume 8, Kroschwitz, J. I., Ed.,Wiley: New York, 1993. (b) hydroxycarboxylic acids, aromatic." Ritzer, E., and Sundermann, R. P. 510. In *Ullmann's Encyclopedia of Industrial Chemistry*; Wiley: Volume A 13, New York.
12. Szmant, H. H. P. 467. *Organic Building Blocks of the Chemical Industry*. Wiley. New York, 1989.
13. Kirsch, M. A., Williams, D.J. *CHEMTECH* 1994. "Understanding the thermoplastic polyester business." *24*(4).
14. (a) Erickson, S.H. P.500 in *Kirk-Othmer Encyclopedia of Chemical Technology*, 3rd Edition, Volume 20, Grayson, M. Ed., Wiley: New York, 1982.
15. Szmant, H. H. *Organic Building Blocks of the Chemical Industry:* Wiley: New York, 1989.
16. Krumenacker, L., Constantini, M. , Pontal, P., and Sentenac, J. P. 996 in *Kirk-Othmer Encyclopedia of Chemical Technology*, Fourth Edition ,Volume. 13. Kroschwitz, J. I.,and Howe-Grant, M., Eds. Wiley: New York, 1995.
17. Varagnat, J. P. 39 in *Kirk-Othmer Encyclopedia of Chemical Technology*, Third Edition, Volume 13, Grayson, M. Ed., Wiley: New York, 1981.
18. Franck, H.-G., and Stadelhofer, J. W. *Industrial Aromatic Chemistry*. Springer-Verlag: New York, 1988.
19. Draths, K. M., Ward, T. L., and Frost, J. W. *J. Am. Chem. Soc.* *114*, 9727, 1992.
20. Davis, D. D., and Kemp, D. R. Pp 466-493 in *Kirk-Othmer Encyclopedia of Chemical Technology*, Fourth Edition, Volume 1. Kroschwitz, J. I., and Howe-Grant, M., Eds., Wiley: New York, 1991.
21. (a) Sato, K., Aoki, M., and Noyori, R. *Science 281*, 1646, 1998. (b) Deng, Y., Ma, Z., Wang, K., and Chen, J. *Green Chem. 1*, 275, 1999. (c) Dugal, M., Sankar, G., Raja, R. and Thomas, J. M. *Angew. Chem. Int. Ed. 39*, 2310, 2000. (d) Raja, R., Sankar, G., and Thomas, J. M. *Angew. Chem. Int. Ed. 39*, 2313, 2000. (e) Besson, M., Blackburn, A., Gallezot, P., Kozynchenko, O., Pigamo, A., and Tennison, S. *Topics Catalysis*, *13*, 253, 2000.
22. (a) Thiemens, M. H., and Trogler, W. C. *Science 251*, 932, 1991. (b) Dickinson, R. E., and Cicerone, R. J. *Nature 319*, 109, 1986.
23. Draths, K. M., and Frost, J. W. *J. Am. Chem. Soc. 116*, 399, 1994.

DISCUSSION

Leo Manzer, DuPont: In my experience working with biologists at DuPont and other places, I have found that they sometimes treat the chemistry too simply. They "cheat," somewhat and equate acids with their salts. In your process, you mentioned, adipate, not adipic acid. I am curious as to whether you are in fact, making the sodium salts, by doing the biology in basic media, or are you actually making the acids?

John Frost: There are several critical impediments that have to be overcome in order to make the process work. The first is catalyst immortalization, so that you can go from batch production to continuous production. This is widely accessible. There is research and development going on in that area.

The salts issue is the second hurdle. Ideally, you want bacteria that operate at a pH of 3, but once you start factoring in the salts, it becomes difficult. The presence of the acid removes the need for sterilization, so you do get something back from it. The salt stream issue remains one of the major impediments, and more work needs to be done to resolve it. It is an issue that people who work in this area don't always appreciate.

Tom Baker, Los Alamos National Laboratory: You commented that you are using bioengineering instead of organometallic compounds to do the catalysis. For catalysis in water, there have been developments using water-soluble compounds as catalysts, such as soluble metal phosphine compounds. One characteristic they have is that they are not very susceptible to the kinds of sulfur compound poisoning that you mentioned. Have you considered using some of these water-soluble catalysts for any of the chemical processing?

John Frost: The way we operate in my group is that, essentially, everybody acts as an organic chemist. We will independently synthesize a critical intermediate and evaluate the chemical details first. Once we get that, then we do the genetic engineering.

The biocatalysis we use to interface is not ideal. There are tremendous opportunities in this area, but we just have not had time to investigate them thoroughly yet.

Rosemarie Szostak, Department of the Army: I have been fascinated with the area of genetically modified organisms (GMOs) for a while, even though I am an inorganic chemist. What regulations are you subject to in your use of these GMOs, and do you envision adverse public opinion about what you are doing?

John Frost: Regulatory hurdles have caused enormous frustration. In this research, we work with a form of *E. coli* called K12. We work in K12 because it is the easiest one for us to get regulatory clearance, both in the United States through the Environmental Protection Agency and in Europe through the European Union.

E. coli may or may not be the best catalyst for making a chemical, but it comes down to what you can get through the regulatory process. There are two types of *E. coli* that people have historically worked with—K12 and another type called B. They were going through the regulatory process back in the days when I was in graduate school in the 1970s. Many people were concerned about "Andromeda strains" in Cambridge, and at the time, the public discussions that happened were extremely important. A lot of validation had to be done, and through the regulatory process, K12 was accepted.

Getting a process with *E. coli B* cleared has been much more difficult. . There are tanks with *E. coli B*, and you will have a hard time even letting people use those tanks. It is also problematic beyond that, unless you get significantly away from the generally recognized as safe (GRAS) organisms.

In Europe, it is very bizarre. For example, Italy's policy has changed overnight. It used to be that whatever we wanted made could be done in Italian labs without much problem. Now it's not as easy. We sent a strain off for production in Italy accompanied by 12 pages of very detailed documentation that included the entire trail, where every gene came from.

In Switzerland, an adamantly anti-biotech country, they were going to ban all recombinant work. I think this was voted down last year. Yet, oddly enough, in Switzerland, we don't have a problem. Going into Switzerland is very easy for us.

In terms of the public perspective, I am lucky. The public has a perception, courtesy of the pharmaceutical industry, that there is indeed a benefit from a GMO. Also, the results of this work are not for human consumption. That protects me from a certain group of people.

However, I have transgenic plant colleagues in the center that I direct. Some of these people have had dire problems with those opposed to their research. We've had two buildings burned down by environmentalists, the second one just last year. This group could be considered a "fringe" group. It gets speaker lists from events like this and starts sending threats by e-mail. We have a detective in our police force that deals with the e-mail threats. They go through a database with the Federal Bureau of Investigation to determine whether it is a credible threat. These labs are guarded by police at night. There is a big difference between a microbe in this context and a transgenic plant that is in the outside world. For the latter, there is more concern about mobilized genetic elements and the issue of food.

Regulatory hurdles are a serious consideration, no matter what you are doing, but to address your question, the problems depend on where you are and what you are doing. Once we start moving into chimeric and mosaic organisms, which is ultimately going to happen when we start going to chromosomal shuffling, it is going to be even more difficult than now.

Alex Bell, University of California at Berkeley: Toward the end of your talk, you alluded to the interplay between biocatalysis and conventional catalysis. I think that is an interface where some exciting opportunities exist. Could you tell us more about what you think are the research needs and opportunities in this area?

John Frost: Let's start with the basic. The problem with the Williamson ether synthesis is that you get a mole of salt per mole of product. When you look at large cellulose processes—say, methylcellulose—you are making a huge amount of salt when you make a huge amount of product, because to get this cellulose, you have to use a caustic of some kind to solubilize the cellulose, which also hydrolyzes most of the alkylating agent. If you did the same process biocatalytically, you could get rid of that salt, as long as you had the microbe capable of doing the alkylation.

A different interface is catalytic upgrading of fermentation broth, which uses both catalytic reduction and catalytic oxidation. For example, let's take a basic one that ought to be easy to do—reduction of a carboxylic acid to a primary alcohol. If you are doing it in a target-oriented, academic environment, you would use diborane. This has quantitative yields. The problem with that is that the stoichiometric reaction comes from boron trifluoride, I believe. That is where the diborane comes from. If you try to do the same thing with catalytic hydrogenation, you start seeing use of copper chromite and nickel catalysts. You can also do this on rhodium now.

When converting a carboxylic acid to the alcohol, the process typically runs at between 100 and 200°C and at 3,000 to 5,000 pounds per square inch hydrogen. So I consider that an example of a very basic organic reaction. The problem is that under those conditions, other chemical reactions, such as hydrocarbon cracking, will occur. It would be wonderful if we had catalytic options that would work at much more mild conditions. Oxidation of organic chemicals is an example of another research area with opportunities. Selective oxidation of hydroxyl groups is an area that has some wonderful opportunities in it.

Panel Discussion

Alex Bell, University of California at Berkeley: Before we start, I would like to suggest thinking about the policy issues involved with the management of carbon that came up in discussions yesterday and have been touched upon again this morning. The question is, How do we translate this area into a research agenda, particularly for the chemical sciences? I think we have an opportunity, with the four gentlemen here on the panel, to address some of these questions.

Richard Alkire, University of Illinois: I would like to ask a question that involves the pace of innovation. I think all the talks today, and many of the others, have proposed new and innovative ways to do process chemistry and processing, which have to be evaluated. I think Leo mentioned that in some cases, corporations can invest 10 or 15 years to do this. So the question is, How do we speed the pace of innovations so that it takes 10 months or, if we do already know the research, 10 weeks, instead of 10 years?

Corporations that make airplanes and cars have figured out how to do this by the strategic use of information technology. They can model fluid flow, heat transfer, and vibration, and then take this and develop lift, drag, weight, and performance characteristics. They can test airplanes before they build them. They can manufacture, assemble, and do inventory control with a high level of information resources.

In the chemistry and chemical engineering area, we can model at the molecular level. We can do heat transport and reaction, optimization, process control. So the tools seem to be available. Yet it would appear that they are not actually used strategically to speed the innovation. The question is, Can information technology be used more strategically than it is by the chemical process industry? Does carbon management give an entry or a driver for its better use? If so, what are the characteristics of applications where a breakthrough or an advancement in that area might first appear.

Patrick Gruber, Cargill Dow LLC: That kind of information actually is used, and we do it, too, each time that we develop a new process, and we have done lots of them. We use all modeling techniques that are available, information-based techniques that are based on fundamentals. Each time we use models, then we stand back and consider what risk we are taking. In the airplane industry, for example, they are working with extremely well-characterized materials, not forefront materials. They are optimizing manufacturing systems based on what they have done before.

In the chemical industry and in the agricultural industry—we have worked in both—the optimization of the distribution systems has been done. The optimization of processes is being modeled. They are well optimized. As to what happens when you try to shorten the development cycle, I can give you a specific example—the scaling up of polylactic acid (PLA).

You do all the chemistry, you do all the fundamental tests, you think you know it. You talk with all the best modeling experts in the world and all the best polymer processing people. Then you say, here is how it should work. Guess what? PLA doesn't behave like other polymers. It has got its little nuances. The only way to find this out is to have put it under those conditions and thus advance the forefront of knowledge in the way this PLA macromolecule behaves in that regime. The knowledge didn't exist before. This happened to us in trying to figure out how to remove the residual lactide monomer effectively.

Everybody was wrong. Fundamentals predicted the wrong thing. It didn't work. Now, from an industrial opportunistic point of view, you have to appreciate that we know how to do it and others don't. So we are happy. That is the problem you run into. If we had just said, for sure it would work—but didn't test it even though everybody who is an expert said it will work—we would have a big problem.

Leo Manzer, DuPont: Let me add a comment there as well. I think how to increase the pace of innovation is a great question to ask. Every time we try to scale up something we get asked the same question. Do you really want to spend another $10 million for a pilot plant? Can we do it any faster?

I have a lot of scars from attempts where we have tried to bypass those kinds of expensive intermediate steps, Richard. Most of the time, it doesn't work. One process in particular, we scaled it up and we found that after a couple of months on-line, we ended up with an impurity in the recycle stream that we never saw in the laboratory work. It appeared only after many months on-line and it couldn't be separated from the final product. It cost all kinds of money to get that out after the plant was built.

I think it is a great question to ask and it is a good objective from an engineering and a modeling research standpoint, but there is a lot of risk in skipping some of those intermediate steps.

Harold Kung: I think I will discuss your question from a somewhat different point of view. You asked, "What does it take to bring innovation to a different level?" I am talking from the experience I learned from a program that was discussed here earlier, the so-called Partnership for a New Generation of Vehicles (PNGV). It was a very successful program in my opinion, and it started with a new paradigm. The industry and government buy into it saying, "We want to build a car that can have 80 miles per gallon fuel economy, with all the other usual attributes."

If the chemical or the fuel industry started with something like that, instead of asking, How can we improve our process by 1% here, 2% here—change the paradigm completely? You must be willing to invest the resources to do it.

A number I heard last year, in one of the National Research Council reports on this PNGV program, is that the companies and government together were spending close to $1 billion in research and development in that program per year. If the chemicals and fuel industries are willing to do something of that sort, they or the researchers can start looking at creative innovations that may or may not pan out. Pieces of those innovations would find their way into improving the existing processes for the industry as a whole.

It is very difficult to schedule innovation, but by putting resources and commitment to it and getting people interested in improving the environment, there will be a lot of good things that can come out.

Patrick Gruber: There is one other aspect I should mention. When we look at a new process, we have a rudimentary idea and we turn it into some kind of process flow diagram and then make a model of it. Then we ask ourselves, "What does it mean in terms of energy and cost? What is it sensitive to?" We don't care so much about the absolute of what it is. We do sensitive analysis and determine which risks we are talking about and where we should focus our effort. That is a slightly different discipline than straight chemical engineering, because we are doing stuff that is total fantasy.

For example, we have all kinds of processes. I bet we have 10 different processes on how to deal with salt in fermentations, sitting there, all of which we predict should work. The question is, Which one do we pilot and prove?

Tom Rauchfuss, University of Illinois: I want to start off with a comment about educational trends that we are seeing at the university level. That is, the Department of Energy (DOE) and the National Science Foundation (NSF), which are fundamental to driving the research that underpins much of what we are talking about today and yesterday, are peanuts compared to the National Institutes of Health (NIH).

NIH is really driving the agenda at the universities, at least in the chemistry departments. The level of funding is perhaps triple what it is in DOE and NSF. Consequently, colleagues, new colleagues, are moving into those areas. Organometallic and process engineering are viewed as almost passé.

This is a major structural change. What we are hearing at the same time is that the basic organometallic and process engineering is critical to what is happening or what is needed in the carbon management area. Funding inequity is a very serious problem, or at least a phenomenon. I don't know whether it is a problem.

To a certain extent, "bio-anything" is viewed by new graduate students and new faculty as extremely sexy, and they are rewarded immediately when they go to Washington, D.C., with four- and five-year grants with direct costs that are hundreds of thousands of dollars more than you are getting from the other agencies.

It is an interesting cultural trend. I am not sure that industry is really prepared for the shortfall of people trained in the correct areas that is coming in a few years. I am interested in hearing what trends you see. For example, John, you probably have some overlap with these areas as well. I would be interested in hearing what you have to say about how this shortfall should be addressed in the National Academies.

John Frost, Michigan State University: I play alphabet soup. My best grants are the ones that are rejected. You know, what you have said hits it right on the head. I will give you an example of what happened to me. I was an associate professor and I was having funding problems. I had just moved, pretty much, into this area. I was looking at the fact that I wasn't going to be able to meet payroll in August. I turned around and I licensed one of my key pieces of technology to a company. That was the single most stupid thing I have done in my life. Then after that, the funding started to go back up. I felt, you know—the first agency to fund me was the Environmental Protection Agency (EPA). It was not NIH. It was not NSF. I hold no grudges. They fund me now.

In terms of the comment you made, I really feel that keenly. When I sit on NIH study sections, I fundamentally wonder what the level of chemical understanding is in the proposals that I see. Some of this is "sexy," but it borders on intellectual masturbation. What can you do on this? NIH has done one thing that is relevant to this area. It is desperately trying to deal with this metabolic engineering issue. Basically, what is going on here is that, Americans are very good at discovering things and then the Japanese eat our lunch. They have not been able to get a study section together on this, which would help chemical engineering. That is because of "insiding." I mean, the last thing people want is another study section that can reduce their piece of the pie. NIH has struggled and started to move somewhat successfully into some areas that are not typically associated with it. I think that your point is well taken, because you also see this in graduate students, in terms of how good their training is. It affects the pace at which we can do research. I don't have any answers. I see the same thing and I play all sides of the coin: United States Department of Agriculture, Environmental Protection Agency, National Science Foundation, National Institutes of Health, whatever else is out there.

What you said is exactly right. One NIH grant is worth three NSF grants. Is that good or bad? I think it is cause for concern. I don't know if that answers your question. Even what I have done, which falls under the "bio-thing," I have felt the hounds nipping at my heels, as it were.

Richard Wool, University of Delaware: I have a question for the panel regarding bio based materials. In terms of making bio-based materials, either you can derive your starting molecules from microbes or you can get your starting molecules from plants, and you can genetically engineer both of those pathways.

Then you can have mixed pathways. From your perspective, in terms of carbon management, are there any strong driving forces in either the molecules from microbes or the molecules from plants that you think should be pursued or combined or mitigated? Do you know where the carbon management—the CO_2 recovery optimum—would be within those systems?

Patrick Gruber: I think for almost all the processes I can imagine, the preferred route will be directly related to the amount of energy used.

Next would be the equivalent of the yield, starting with the right oxidation states. One of the things that I have seen people do is take glucose, which is highly oxidized, then do a bunch of chemistry to it and turn it back into something that has no oxygen at all. This is not a good idea. There has got to be some more fundamental knowledge gained in picking the right raw materials, and manipulating them.

This sounds quite simple. It isn't, though. When you start applying biotechnology techniques in plants or microbes or whatever—and John showed a good example—where they are doing one that has a couple of double bonds. I can imagine a scenario where you have manipulated something to make a new raw material. It then is a totally different game and paradigm. You have to be thinking up front about where that optimum is, what it looks like, how might it come to be, and then make trade-offs to actually reach it.

In general, if you have maintained your yield and minimized your energy, you will usually have done well on minimizing carbon dioxide production.

Richard Wool: I wonder if Leo would care to comment on DuPont's perspective on this.

Leo Manzer: I can't comment. I don't know. I would make a comment, though, totally unrelated, but I want to emphasize a point that Pat and John made as well. I think there are real opportunities here, as we talk about renewable resources and so on. It is not to try to do it all with enzymes or bugs or whatever or chemically. There is a great opportunity here—and I am involved in this myself—in trying to learn some of the biology so I can talk with some of these guys.

Sitting through some of these conferences is pretty tough when they go with acronyms all over the place, but there is an opportunity to say, "Hey, let's take it part way and then do what I ought to do best, and that is chemical catalysis and take it the rest of the way." I think that is where the future lies, certainly in DuPont.

Richard Wool: That was my last question. This creates a beautiful interdisciplinary interface that rarely exists, even within universities, where it should happen the easiest. Within industry, these units tend to be very well accepted.

Leo Manzer: It is in handfuls of people that I am talking about who are superb at cutting across boundaries. It is handfuls of people. It is really tough, because I think it is the way to go. In biology or chemistry, everyone has to understand economics. This means they have to know about the process technology.

The chemical engineers that are doing it need to understand both the chemistry and the biological issues that arise, as in catalysis or some other type of thing.

We have had problems when we have people who cannot get out of their little box—these are the rules and things that I learned working with petrochemical—based products in the chemical industry. Get rid of those paradigms and let's get some new ones, to the point of getting rid of people.

Richard Wool: Is there a suggestion here for the federal government to begin looking at new centers within universities or within industry, or joint between them, about how to have a mix of the right kind of people to address these problems? Something needs to be done. These people don't exist, as we talk. I don't know what the answer is. I do know it is disturbing if people are sent to make more metabolic engineering tools. I don't like that very much. We need to solve problems and get things done and implemented to make money some way. Tools are nice, if I am going to start my own company and sell a tool, but I'm not. My company makes products. Therefore, this means an integrated system of solving problems using process technologies, chemical engineering, chemistry, and biology, all of it integrated. This is what needs to be done. This is what needs to get funded. It is disappointing if the funding shifts so for over to the bio side that we forget the rest.

Leo Manzer: I wonder if this isn't an opportunity for universities to try to train students in chemistry and biology more on the process side of things. It is just learning the language. It is still chemistry when you get down to the nuts and bolts of the thing. I sat through a meeting a couple of weeks ago on microbiology, and I didn't understand one word. I didn't know whether they were successful or not. It is a matter of taking the time to learn the language. You think it is hard to get chemists and engineers together? That is insignificant compared to getting biologists and chemists together, I think.

David Keith, Carnegie Mellon University: This panel has illustrated wonderfully the enormous potential of chemical and biochemical process engineering to really improve the environmental performance of our energy and chemical technologies. I think the promise is extraordinary.

This panel has also illustrated some real thoughtfulness about the complexity of what environmental performance means. I address a question to the panel but, first, a critique.

The critique is that I feel that many people, when thinking about better chemicals and especially "green" chemical processes, have reasonably naive notions about what is better environmental performance. As a starting point, there is the simple notion that using fossil fuels is always worse than using biochemicals.

So how do we inject into folks like you, who are thinking about designing better processes, the best knowledge we have about what really constitutes environmental insult. How do we measure it and how we do that kind of policy analysis? How do we do this at a process level and how do we do it at the level of national policy?

Patrick Gruber: I will take a shot here. The issue best illustrated is life-cycle inventory (LCI) analysis. In theory, it is quite easy to calculate the energy that we use. If I ask the simple question: Where did that energy come from? And ask what does it mean? This is a problem. The data is old, for one. Electric companies have improved their processes, but these improvements are not reflected in avalanche data.

Understanding the source matters. The standard assumptions for LCI don't differentiate whether the source was clean or not would be, and this was a clean place, because it is all uniform across the grid. With those kinds of assumptions, you are disinclined to have a discussion with the electric company that has the wire going to your plant. I need to be able to say, Hey, my customers care, I need this cleaner. I need to have specific data so that we can have discussion. This is what needs to occur in order to make a dent. So what this comes down to is, How do you change supplier behavior? Without information and data that are real, you can't have the discussion.

The same thing would be true in farming. If we want corn LCI information, we have got to have data. Who is going around talking to farmers to see what farming practices are used? What inputs are you using? Are you improving? How does it roll up at county levels? We need data that is useful for decision-making, then we can determine more fully how to improve.

John Frost: I could add something about educating people on environmental costs and what not, I have seen texts in this area, et cetera. I am underwhelmed by what I see.

David Keith: So am I.

John Frost: It falls into the category of fuzzy. What I think is—and actually I think you alluded to this earlier—if you train people in a traditional sense to focus on cost of carbon, titer, and yield, you will, in that process, address environmental issues. I am not so convinced that we need to be teaching people how to save Bambi. I think we need to focus on our nuts and bolts, and from that, I think good things will happen.

David Keith: I certainly agree we don't need to teach people how to save Bambi. What I want to do is connect hard-headed assessments of the environmental costs of various industrial activities to the folks who are designing those processes. There is a lot of soft-headed routinized calculation of environmental insult that, in fact, doesn't hold up under serious scrutiny.

Patrick Gruber: I think you answered your own question. What we need is to develop data and models based on regional and local data. This is tough, but it needs to be done. Otherwise, we are going to get wrong decisions.

A product like ours gets sold across the world. We need to take into account local waste disposal systems. For me to assume that it is going to be landfilled in Germany is a mistake. In Germany, disposable plastics are going to be incinerated.

Glen Crosby, Washington State University: I would like to return to one issue that was brought up by Tom Rauchfuss, concerning a very serious imbalance in the amount of funding that is going into health-related or biologically oriented research in the universities and the amount going into physical sciences and engineering.

I think this is really serious. It is a lot more serious than people know. I think there are ways of addressing it. One way is, of course, it is beginning to happen. Harold Varmus, when he was head of NIH, went out of his way to emphasize the importance of having developments in the physical sciences to support the biomedical community. I think that was a very fine statement and I think it had an effect.

The more serious problem, in my opinion, is why students are not going into the chemical and physical sciences. That problem starts farther back. It starts in the high schools in this country because no one now majoring in physics or chemistry intends to go into a high school and teach.

The people who are now teaching physical sciences in high schools are trained mainly in biology. They reflect that attitude and they also reflect an insecurity regarding our sciences when they teach them. So there is a very serious mismatch between the education of teachers and what they are required to teach.

As a member of the board of directors of the American Chemical Society, I have been extremely worried about this issue. I have proposed that we run a professional development program at the national level, funded through industry. We should take those teachers who are educated in the biological sciences, and who are being forced to teach physics and chemistry without solid knowledge in it, and run a professional development program in the molecular sciences. The program would try to build on their biological interest and background, and enable them to teach physics and chemistry better.

Can it be done? Yes. I have done it in my own state. It nearly killed me, but I did it in my own state, while being underfunded for this, but I think it can be done nationally.

If we are going to have an infrastructure in this country that is going to lead us and keep us in the forefront in the twenty-first century, we have to do something about the basis of the scientific learning that occurs in high schools and therefore goes through the community colleges and universities.

It is now slowly eroding. The flight of students into the biological sciences is not necessarily all pull. Some of it is push, because they are not educated in mathematics and physical sciences in high school and they find themselves unable to study these things in colleges and universities. They are unable to do the very simple things that are necessary to be successful in those elementary courses because of an extremely poor background in high school. We are going to have to address this issue. It is not just something that we can throw away, and it is going to have to be more than just government programs.

It is going to have to be industrial input. As I listened today, I thought to myself, "if some of the things I heard today were being talked about in freshman chemistry classes, it would be extremely exciting to kids."

So the message here is that we cannot have the kinds of people you want emerging from our graduate programs unless we give serious thought to what is happening in our high schools, particularly, and what is happening in our school systems. We cannot expect teachers, people who become highly educated, to step into our schools and be paid $22,000 a year. The American tax payer is going to have to start paying. Otherwise, we will have to import all of our serious scientific and engineering personnel from abroad. This is something that we are going to have to face as a nation.

Klaus Lackner, Los Alamos National Laboratory: It appears, from presentations over the last two days, that the overarching opinion seems to be that carbon emission will be constrained. Somehow or another we should use less of it, particularly if it is of a fossil nature, and we have to put away what we use.

If you believe what is going on in the sequestration world, nobody believes this will be a success unless the price per ton of carbon is less than $100. The impact on the industry is similar to raising the price of oil by roughly $10 a barrel. We saw this size impact last year and survived. Clearly, you want to deal with this and improve all of these issues, but I think you have to keep the orders of magnitude when it comes to the environmental issues.

We are talking a lot about what to do technically and scientifically in these areas where we can make improvements. Clearly you want them and they are very worthwhile. Ultimately, chemistry will be needed in dealing with the issues of dealing with CO_2. We need to have a better understanding of CO_2 chemistry. If you put it underground, we have to deal with the geochemistry of how it interacts with where it goes. If we form mineral carbonates, we are clearly constrained by a lack of understanding of how to do this. More work in this area by many more people will be required to make these processes elegant and efficient.

I really would try to put a plea in for the chemistry field as a whole, to take on the big challenge that needs to be dealt with if you want to use fossil fuel for the next 50 years. Given the time constants of changing infrastructure, we will have to deal with the sequestration issue and do it right.

Jae had it absolutely right. We have to go to zero emission, which is a nontrivial effort. I view all of this as being very important and actually driven by economic constraints. If the cost of carbon goes up, all of the things discussed earlier are important, but let's not forget the biggest driver in it—which is the combustion of fossil fuels.

John Frost: If you look at nature in terms of CO_2 removal capability, there are ribulose bisphosphate carboxylase, and phosphoenolpyruvate carboxylase, which are marvelous. Ribulose bisphosphate carboxylase is less effective because oxygen is required in its reaction with CO_2. Thus, in nature, there are enormous opportunities for CO_2 sequestration.

Klaus Lackner: I absolutely agree with that. I just would say, let's have some of the focus on those areas.

John Frost: I think that will happen naturally as you go along. It is also incumbent on chemists to understand that nature does it a lot better than they do. They need to study these systems to see if, in a truly chemical sense, they are able to create the chemical systems that can sequester CO_2, or improve on those systems that are doing it.

So, this goes back to something that was addressed earlier in terms of ignorance between various fields. I mean, it crosses everything. I am ignorant about a lot of things. Other people are ignorant about a lot of things. CO_2 fixation in large-scale fermentors is doable at 1 atmosphere. You don't necessarily have to clean gases up. Nitrogen and sulfur are fine for bugs. What you want to do is genetically engineer your organism so that, with proper weathering, it would become appropriate humus. I don't have the foggiest notion of what direction to genetically engineer that microbe. I basically have to talk to Dr. Dirt and I don't know where Dr. Dirt is.

That is one of these field issues. I think there are enormous opportunities that are available in chemistry. There are also some opportunities that we haven't even talked about. For example, one of the things you notice in aerobics, is that you produce about a gram of biomass per gram of product. It is not until you go anaerobic are you in good shape.

Klaus Lackner: When I phrased my question, I didn't want to exclude anything—quite the contrary. We had a lot of discussion on the policy level about these hundreds to thousands of billions of tons of carbon we will, over the next century, have to deal with. I think we should focus some of the energy across the board on that question.

John Frost: Let me make one other comment. In some grant applications, you have to write an environmental impact statement. These things are usually toss-offs that aren't taken seriously. It would be useful if they were taken seriously. It would absolutely force people to start talking to people in other fields. To give a classic example, a very good transgenic plant person received a very large grant to do functional genomics in hemicellulose biosynthesis. I heard this and I went over and started talking, "Gee, this is great, this is great, this is great." I realized that he didn't have the foggiest understanding of why you would be interested in a pentose stream. It finally came back that he didn't have the foggiest notion of what a theoretical yield was in microbe-catalyzed conversion for use of a pentose as a starting material versus use of a hexose as a starting material.

I realized that this wasn't even necessary in the proposal process. In the review process, it wasn't even deemed important. If you have that kind of activity, I think you would have people stepping up to the plate in the fashion that you are saying you wish people would.

Patrick Gruber: Oddly enough, when you said we need some catalytic means to do sequestration, it wasn't until after a while that I considered nonbiological means. It is just chemistry done differently. It doesn't matter which it is. If that is a valuable problem to solve, apply the tools and solve it. We need to figure out what catalyst, chemistry, and systems and process engineering.

Digby MacDonald, Pennsylvania State University: This question is for the two gentlemen from industry. In these types of panels, we frequently hear about what industry needs in students in view of research. What do you do to support research or students, and what do you plan to do in the future to increase whatever programs you have going on.

Leo Manzer: At DuPont, for example, to help students, we donate a lot of money to departments across the board. Maybe the shift is a little bit away from traditional areas into newer areas, but there is a lot of money going to support students, research projects, and so on in those areas that we see as critical to the future. Maybe this doesn't sit well with some areas that are not perceived to be exciting anymore, but that is just the way it is. We have students working in the laboratories for the summer.

Digby MacDonald: Could you comment on any increase in the future in this support and which areas you plan to support?

Leo Manzer: I think it is pretty clear, if you read the press, that our future is in the area of life sciences. Therefore, it is only fair to assume that money is going to go into those types of areas and away from more traditional chemistry and chemical engineering areas.

I may or may not agree with this, but that is the way it is. If this is the future, you might as well go with it. This is why I have decided to move myself over into the biological area. As painful as it may be, I am trying to talk to these folks.

What I find very, very interesting is that they haven't got a clue of what is going on in the world. The biology folks can splice genes and they can do all kinds of really nifty stuff. I spent three weeks in a lab doing a lot of pipetting and decided doing it once was enough. I learned a little bit of the words, but they don't know anything about process development. Making salts is fine with them, but I have been trying to solve these problems for years, and recycling salts just doesn't make it when you are dealing with cheap things like adipic acid. So I think there is an opportunity for the areas to get together and for an educational process to proceed both ways.

Patrick Gruber: Cargill Dow is a small company. So your question is probably better answered by either Cargill or Dow. We have our select universities that we fund, work with, and hire students. We give them money because we want them trained a certain way. For us, it is not huge amounts of money, but it is enough to make a dent in their programs. We also have summer programs and internships. All the time we are trying to find people we would like to hire.

C. B. Panchal, Argonne National Laboratory: I would like to go back two issues. Yesterday we talked about energy and today we are talking about chemicals with selective reactions of the bioproducts. In the United States and worldwide, we should identify 10 of the basic chemicals we need, such as styrene. Then, alternate chemical paths to make them should be developed.

That will be good in this way. Industry is going to continue. It is going to improve. Much less energy is required today than 10 or 15 years ago per unit produced. What we need is the development of alternative chemical paths or alternative processes. If you look at the amount of solvent we use in industry and at the consumer level, it is a tremendous amount and we don't know what happens at the end. Even if only 10% escapes, you can see the impact on the environment.

Look at the total picture, not just productivity, but side reaction suppression, which is the major cost for purification and separation. Let's find a way that does not require too much separation, like membrane reactors.

I don't know that you need to comment on that one. What we are looking for is a state of change. We are not discovering new chemicals. It is the same chemical made in different ways.

Leo Manzer: I think there is an opportunity to do both things. Certainly looking at new chemical routes to the top 10 or 15 chemicals is fine. I think the other way to go is look at the properties that we are after and go after alternates—polylactic acid, for example, instead of styrene. Instead of building these big massive plants that consume lots of energy and fossil fuels, find some other way to get the desired properties from a different product altogether.

Another comment, too, on solvents—just keep in mind, once you have taken water out of the tap and put it into a process, it isn't water any more. It needs to be purified and treated like anything else.

David Cole, Oak Ridge National Laboratory: I just want to make a comment to echo what Lackner mentioned a minute ago.

The problem we are dealing with is a huge amount of CO_2 going into the atmosphere. As many of the multilab reports have pointed out, there is going to have to be a multitude of technologies used to address the problem, both from the chemical processing point of view and from the geologic point of view. We need to look at all those various technologies to address this problem.

We are talking about mitigating CO_2 from the power plant we have near Oak Ridge, the Bovon power plant, which produces tons of CO_2 per day. There was a fair bit of activity afoot using microbes that respire or take up CO_2 and make carbonate minerals, as well as algae. They do this fairly quantitatively. This is an area of active research. I think it would be good to get that out in the open so people know about it. Of course, the issue of genetic engineering to optimize the carbonate precipitation process of these microbes also is something of interest and is being worked.

The specific question I have for Dr. Frost is, With regard to *Escherichia coli,* are there other microbes that people are looking at across the board to produce chemicals in the way that you have described, or is *E. coli* the only microbe? Are there other opportunities out there to look at various reaction pathways that depend on the microbe's starting point?

John Frost: In the United States, you have a lot of K12 processes. In Japan, it is *Corynebacterium* spp. Another microbe *Gluconobacter oxydans* is used in the manufacture of ascorbic acid.

I think what you are asking is, What is the best organism to use and how does the best organism relate to some of these issues? I think that if you are starting to define the best organism, you are looking for an organism that uses—you can define some of these things pretty well. You want facilitated diffusion for glucose uptake so that you are not sacrificing yield for rates of glucose uptake because you can always have plenty of glucose in a tank. You want an organism that works very effectively at low pH. *Gluconobacter*, which is related to *Acetobacter*, which goes back to Louis Pasteur, is an awesome example. These organisms synthesize 0.4 moles of acetic acid, and are very happy in 0.4 M acetic acid. These types of acidic bacteria have been used not that far back. They are useful organisms.

The other thing that you want is an organism that naturally expresses CO_2-fixing capacity. I put my money on pyruvate carboxylase because it requires adenosine triphosphate (ATP) versus phosphoenolpyruvate (PEP) and it doesn't have oxygenase reactivity.

Now, where things become very interesting concerns the best reduction capability. The best reduction capability is in the anaerobes. If you try to use an anaerobe as a production platform, it grows glacially and at very low density, which presents problems. So you start getting into the interesting and potentially controversial area of creating chimeric organisms. You take the desired megabase out of an anaerobe, de novo resynthesize it for proper codon usage and proper folding, and introduce these into a *Klebsiella*, *E. Coli*, or *Gluconobacter* backbone to create the catalyst that you need. Everything that you need is out there, but it does not exist in a single organism. Even if it were to exist in a single organism, you have some very serious regulatory issues right now as to whether you would be able to use that organism.

David Cole: That was where I was heading. I was also simply trying to point out that there is a multitude of natural microbe species that, from an ecological point of view, may be used in a simpler processes, compared to complex chemistry.

These bacteria actually fix CO_2 directly and dump out a biomineral. They are not as efficient as direct chemistry but nonetheless provide an interesting pathway for fixing carbon in a bulk kind of way.

John Frost: Your comment about the CO_2-fixing carbonate bacteria is a really neat area that has a lot of potential. This also gets back to what I was suggesting in the biological area, which is somewhat related to what we have been talking about in chemistry.

When you get away from *Streptomyces*, *E. coli*, *Arabidopsis*, and you get into some of these other areas, there aren't techniques to transform these organisms, nor are there people who can get money to do it. This is hard stuff. For the *Streptomyces*, it is my country for a good transformation technique.

In these atypical organisms that you are talking about, you have to go back to the 1970s in *E. coli*, and create a whole area of fundamental underpinnings that is going to allow you to manipulate them. There are no funding agencies doing that. It is a real problem. We have become really narrow in terms of what we can manipulate.

David Cole: I agree, and to make one last comment, I think the problem is this disconnect between different disciplines. We talk microbiology community versus geochemistry community versus chemical community. Without a connection between these various disciplinary approaches, we are going to act as separate entities and we are not going to have the connection we need to solve some of these problems.

To do genetic engineering on a natural microbe, such as an epithermophilic or mesophilic bacteria we find underground, is almost impossible. No one wants to talk to us about that kind of issue. They only want to see us use it as a means to remediate a contaminated zone. We have got to go back to a point where we can start understanding the chemical and metabolic processes of these kinds of microbes before we can even begin to attempt to do the fixation. Yet they are out there. That is the key issue.

Closing Remarks.

Tobin Marks, Northwestern University: Let me say a few things as the co-organizer. First of all, I find that I have come away tremendously stimulated. I have learned a great deal. I have gained a new perspective on how important the carbon management issue it is, how complex it is, how difficult it is to discern green versus non-green, and what the life-cycle analyses are. I have learned a great many things.

I came away also with the feeling that there really is a need for a very broad agenda—one that is molecularly, macromolecularly, and biologically oriented. I think there is a need for mechanisms that will stimulate interdisciplinary research, maybe through centers, but also stimulate interdisciplinary education in terms of preparing a work force that can attack these kinds of problems. So this is one thing that I thought was really terrific about this workshop, and we sincerely thank all of the speakers for their contributions.

Appendixes

A

Workshop Participants

Justine Alchowiak, U.S. Department of Energy
Richard C. Alkire, University of Illinois
Gary C. April, University of Alabama
Richard A. Bajura, West Virginia University
R. Thomas Baker, Los Alamos National Laboratory
Patricia A. Baisden, Lawrence Livermore National Laboratory
Krishnan Balasubramanian, University of California, Davis
Alexis T. Bell, University of California, Berkeley
Adolph Beyerlein, Clemson University
Robert Bloksberg-Fireovid, National Institute of Standards and Technology
David C. Bonner, Rohm and Haas Company
Thomas F. Brownscombe, Shell Chemical Company
Donald M. Burland, National Science Foundation
Richard Cavanagh, National Institute of Standards and Technology
Margaret Cavanaugh, National Science Foundation
Thomas W. Chapman, National Science Foundation
David Chock, Ford Motor Company
Steven S.C. Chuang, The University of Akron
David Cole, Oak Ridge National Laboratory
Walter G. Copan, The Lubrizol Corporation
Geraldine V. Cox, EUROTECH, Ltd.
Carol A. Creutz, Brookhaven National Laboratory
Glenn A. Crosby, Washington State University
Daniel L. Dubois, National Renewable Energy Laboratory
Stan A. Duraj, Cleveland State University
James A. Edmonds, Pacific Northwest National Laboratory
Gary Epling, University of Connecticut

Giuseppe Fedegari, Fedegari Autoclavi SpA
Frederick P. Fendt, Rohm and Haas Company
Farley Fisher, National Science Foundation
Brian P. Flannery, Exxon Mobil Corporation
Emory A. Ford, Equistar Petrochemical Co.
Richard D. Foust, Jr. Northern Arizona University
Hans Friedericy, Honeywell, Inc.
John W. Frost, Michigan State University
Etsuko Fujita, Brookhaven National Laboratory
Jean H. Futrell, Pacific Northwest National Laboratory
Dorothy H. Gibson, University of Louisville
Patrick R. Gruber, Cargill Dow LLC
J.Woods Halley, Department of Physics
Heinz Heinemann, Lawrence Berkeley National Laboratory
George R. Helz, University of Maryland
Jason Hitchcock, U.S. Department of Agriculture
Jerry E. Hunt, Argonne National Laboratory
Andrew Kaldor, Exxon Mobil
David W. Keith, Carnegie Mellon University
Dahv Kliner, Sandia National Laboratory
Harold H. Kung, Northwestern University
Klaus S. Lackner, Los Alamos National Laboratory
Antonio O. Lau, BP Chemicals
Dennis L. Lichtenberger, University of Arizona
Michael J. Lockett, Praxair Inc.
Keith E. Lucas, Naval Research Laboratory
Digby D. MacDonald, Pennsylvania State University
Leo E. Manzer, E.I. du Pont de Nemours & Company
Robert S. Marianelli, Office of Science and Technology Policy
Tobin J. Marks, Northwestern University
William S. Millman, U.S. Department of Energy
James A. Moore, Rensselaer Polytechnic Institute
Dennis F. Naugle, Research Triangle Institute
Chandrakant B. Panchal, Argonne National Laboratory
Charles Peden, Pacific Northwest National Laboratory
Thomas B. Rauchfuss, University of Illinois
Douglas Ray, Pacific Northwest National Laboratory
Mark W. Renner, Brookhaven National Laboratory
D. Paul Rillema, Wichita State University
Sharon M. Robinson, Oak Ridge National Laboratory
Howard Saltsburg, Tufts University
Eric A. Schmieman, Pacific Northwest National Laboratory
Peter Schultz, National Research Council
Darlene Schuster, American Institute of Chemical Engineers
Jeffrey J. Siirola, Eastman Chemical Company
Aleksandar Slavejkov, Air Products & Chemicals, Inc.

APPENDIX A

Christine S. Sloane, General Motors
Jack Solomon, Praxair Inc.
James A. Spearot, General Motors
Peter G. Stansberry, West Virginia University
Ellen B. Stechel, Ford Motor Company
John Stringer, Electric Power Research Institute
Kyung W. Suh, The Dow Chemical Company
Rosemarie Szostak, U.S. Department of the Army
David C. Thomas, BP Amoco Corporation
John A.Turner, National Renewable Energy Laboratory
Olaf Walter, Institute for Technical Chemistry
G. Paul Willhite, University of Kansas
Bobby Wilson, Texas Southern University
Robert B. Wilson, Jr., SRI International
Alan M. Wolsky, Argonne National Laboratory
Richard P. Wool, University of Delaware
Staff: Maria P. Jones, Ruth McDiarmid, Sybil A. Paige, Douglas J. Raber

B

Biographical Sketches of Workshop Speakers

Carol Creutz is senior chemist at Brookhaven National Laboratory. Dr. Creutz received her B.S. in chemistry in 1966 from the University of California, Los Angeles, and her Ph.D. 1971 in chemistry from Stanford University. She was an assistant professor at Georgetown University from 1970 to1971 before joining the staff at Brookhaven National Laboratory Chemistry Department, where she served as chair from 1995 to 2000. Her professional activities include service on the Chemistry Research Evaluation Panel for the Air Force Office of Scientific Research (1979-1983); member of the Editorial Board, *Inorganic Chemistry* (1988-1991); councilor, American Chemical Society, Inorganic Division (1992-1995); member of the National Research Council (NRC) Committee on Prudent Practices for Handling, Storage, and Disposal of Chemicals in the Laboratory (1992-1995); and member of the National Research Council Committee on Design, Construction and Renovation of Laboratory Facilities (1998-1999).

Dr. Creutz's research interests include kinetics and mechanisms of ground and excited-state reactions of transition metal complexes, homogeneous catalysis in water, and charge transfer processes in nanoscale clusters.

James A. Edmonds is a chief scientist and technical leader of economic programs at the Pacific Northwest National Laboratory (PNNL). Dr. Edmonds heads an international global change research program at PNNL with active collaborations in more than a dozen institutions and countries. Dr. Edmonds is well known for his contributions to the integrated assessment of climate change—the examination of interactions between energy, technology, policy, and the environment. He has expounded extensively on the subject of global change including books, papers, and presentations. His books on the subject of global change include *Global Energy Assessing the Future*, with John Reilly (Oxford University Press). His book with Don Wuebbles, *A Primer on Greenhouse Gases*, won the scientific book of the year award at the Lawrence Livermore National Laboratory. He presently serves as a lead author for the Intergovernmental Panel on Climate Change (IPCC) third assessment report, currently under way.

Dr. Edmonds's current research focuses on the application of integrated assessment models to the development of a long-term, global energy technology strategy to address climate change. His Global

Climate Change Group at PNNL received the Director's Award for Research Excellence in 1995. In 1997, Dr. Edmonds received the BER50 Award from the U. S. Department of Energy in recognition of his research accomplishments, and he recently received the Stanford Energy Modeling Forum Hall of Fame Award (2000). Dr. Edmonds was trained as an economist with a B.A. from Kalamazoo College (1969) and M.A. (1972) and Ph.D. (1974) from Duke University.

Dr. Brian P. Flannery is science, strategy and programs manager in the Safety, Health and Environment Department, Exxon Mobil Corporation. Before joining Exxon he received degrees in astrophysics from Princeton (B.A. 1970) and from the University of California Santa Cruz (Ph.D. 1974); he was a postdoctoral fellow at the Institute for Advanced Study in Princeton (1974-1976) and was assistant and associate professor at Harvard University (1976-1980). Since joining Corporate Research, Exxon Research and Engineering Company in 1980, Flannery has worked in research, supervisory, and management roles involving theoretical science, mathematical modeling, and the environment. At Exxon he led the effort to develop a new form of microscopy utilizing synchrotron x-ray radiation to produce noninvasive, three-dimensional images of the internal structure of small objects. Flannery is coauthor of the widely used reference *Numerical Recipes: The Art of Scientific Computing*.

Since 1980, Flannery has been involved in research and policy analysis of scientific, technical, economic, and political issues related to global climate change. He served on the *State-of-the-Art Review of Greenhouse Science* of the U.S. Department of Energy 1984-1986, where he coauthored the chapter on transient climate change. He was a member of the *Scientific Advisory Subcommittee on Climate Change* of the U.S. Environmental Protection Agency (1988-1990). He served on the editorial committee of *Annual Reviews of Energy and Environment*, and *Consequences*, and he was a member of the *Evaluation Committee* of the International Geosphere-Biosphere Program. Currently he participates in the *Third Assessment Report* of the IPCC as lead author in Working Group III.

Through the Global Climate Change Working Group of the International Petroleum Industry Environmental Conservation Association, Flannery has organized international seminars, workshops, and symposia that address scientific, technical, social, economic, and policy aspects of global climate change. These include the 1992 Rome symposium, *Global Change: A Petroleum Industry Perspective*, the 1993 Lisbon Experts Workshop *Socio-Economic Assessment of Global Climate Change*, the 1996 Paris symposium *Critical Issues in the Economics of Climate Change*, and the 1999 Milan workshop *Kyoto Mechanisms and Compliance*. On behalf of industry he participates as an observer at meetings of the Intergovernmental Panel on Climate Change and the Framework Convention on Climate Change (FCCC).

John W. Frost is a professor in the Departments of Chemistry and Chemical Engineering and director of the Center for Plant Products and Technologies at Michigan State University. He received his B.S. in chemistry from Purdue University and his Ph.D. from the Massachusetts Institute of Technology (MIT), and was a postdoctoral fellow at Harvard University. The Frost group genetically engineers and uses recombinant microbes as synthetic catalysts and interfaces this type of biocatalysis with chemical catalysis. His research has focused on elaborating microbe-catalyzed syntheses of starting materials critical to the manufacture of pharmaceuticals as a replacement for the current isolation of these starting materials from exotic natural sources. Hoffmann La Roche is currently employing a Frost group microbe commercially to synthesize shikimic acid, which is the starting material used in the manufacture of the anti-influenza drug Tamiflu. Frost group research is also directed toward employing recombinant microbes in syntheses of larger-volume chemicals including adipic acid, catechol, hydroquinone, and vanillin. These microbe-catalyzed syntheses exploit renewable feedstocks (starch, cellulose, hemicellulose) and nontoxic starting materials (glucose, xylose, arabinose, glycerol) as sustainable, environmentally benign alternatives to the nonrenewable feedstocks (petroleum) and toxic starting materials (benzene, toluene) that are currently employed in chemical manufacture. Professor Frost and his wife and collaborator, Professor Karen M. Frost, were awarded the Presidential Green Challenge Award for these research efforts.

Patrick R. Gruber is currently vice president and chief technology officer, Cargill Dow LLC, a joint venture between Cargill, Incorporated, and Dow Chemical. Dr. Gruber began working with Cargill Dow in 1997 and has served as vice president since Cargill Dow's formation. Dr. Gruber joined Cargill Dow full-time beginning in January 2000. During his tenure at Cargill he served in a wide range of roles in the technology and business development area. Dr. Gruber has spent his career developing technology and business opportunities in the area of chemical products made from renewable resources targeted to animal feed products, food ingredients, and industrial chemicals.

Dr. Gruber has served on strategy and business teams at the division level of Cargill. From 1995 to 1998 he was director of technology development for Cargill's bioproducts areas. From 1998 through 1999, he served as technical director of Cargill's BioScience Division where as a member of the Business Management Team, he was involved with identifying and starting up a variety of new businesses, as well as building capability in the food products and animal nutrition area.

Before joining Cargill Dow, Dr. Gruber was president of Lactech, a technology development company that successfully developed lactic acid technology, which was licensed to Cargill, Incorporated. In 1989 he was named leader of Cargill's renewable bioplastics project with responsibility for the development and marketing of a lactic acid polymer as NatureWorks. In this general management role, Gruber led the development from concept through technical and market validation, building the organization that formed the core of Cargill Dow.

Dr. Gruber has 37 U.S. patents issued with more than a dozen pending. In 1998, he received Inventor of the Year Award from Minnesota Patent Lawyers. In 1993, he received *R & D Magazine's* Top 100 Inventions of the Year Award for advances in stabilizing enzymes. Dr. Gruber served as one of the program reviewers of the Department of Energy's (DOE) Biofuels Program (1998 and 1999). He has been counselor of Bio Environmentally Degradable Polymer Society (BEPDS) since 1997.

Gruber received a bachelor's degree from the University of Saint Thomas, St. Paul, Minnesota, in 1983, where he majored in chemistry and biology. He earned a doctorate in chemistry from the University of Minnesota in 1987, and he also has a master's in business administration from the Carslon School of Management at the University of Minnesota (1994).

David W. Keith has been a faculty member in the Department of Engineering and Public Health at Carnegie Mellon University since 1999. His current research centers on the use of fossil fuels without atmospheric emissions of carbon dioxide by means of carbon sequestration. His research aims to understand the economic and regulatory implications of this rapidly evolving technology. Questions range from near-term technology-based cost estimation, to attempts to understand the path dependency of technical evolution—for example, how would entry of carbon management into the electric sector change prospects for hydrogen as a secondary energy carrier? In addition, Dr. Keith's research interests include geoengineering, biomass energy, and the use of quantified expert judgment in policy analysis.

Dr. Keith trained as an experimental physicist at MIT (Ph.D., 1991) where he developed an interferometer for atoms. During 1991-1999 he worked in atmospheric science, first at the National Center for Atmospheric Research (NCAR) and then at Harvard, he also a collaborated in the research program on climate related public policy at Carnegie Mellon as adjunct faculty and as an investigator in the Center for the Human Dimensions of Global Change. As an atmospheric scientist in Professor James Anderson's group at Harvard, Dr. Keith led the development of a new Fourier-transform spectrometer that flies on the National Aeronautics and Space Administration's ER-2 and worked as project scientist on Arrhenius, a proposed satellite aimed at establishing an accurate benchmark of infrared radiance observations for the purpose of detecting climate change. He continues to collaborate on high-accuracy radiance measurements.

Harold H. Kung is professor of chemical engineering at Northwestern University, where his areas of research include surface chemistry, catalysis, and chemical reaction engineering. His professional experience includes work as a research chemist at E.I. du Pont de Nemours & Co., Inc. He is recipient of the P.H. Emmett Award and the Robert Burwell Lectureship Award from the North American Catalysis Society, the Herman Pines Award of the Chicago Catalysis Club, the Japanese Society for the Promotion of Science Fellowship, the John McClanahan Henske Distinguished Lectureship of Yale University, and the Olaf A. Hougen Professorship at the University of Wisconsin, Madison. He is editor of *Applied Catalysis A: General*. He has a Ph.D. in chemistry from Northwestern University.

Leo E. Manzer is a DuPont fellow in DuPont's Central Science and Engineering Laboratories at the Experimental Station in Wilmington, Delaware. He was born and educated in Canada, and after receiving his Ph.D. in chemistry from the University of Western Ontario, Canada, in 1973, he joined DuPont in Wilmington. During his career, he has held a variety of positions in Delaware and Texas, overseeing research programs in homogeneous and heterogeneous catalysis. He founded and directed the Corporate Catalysis Center at DuPont from 1987 to 1993. Dr. Manzer is a member of the North American Catalysis Society and is an adjunct professor in the Departments of Chemical Engineering, Chemistry and Biochemistry at the University of Delaware. He is on the editorial boards of several major catalysis journals and is actively involved in promoting the value of catalysis to society. Dr. Manzer has been involved in all aspects of catalysis in DuPont and led the research effort for the development of alternatives to chlorofluorocarbons.

Dr. Manzer is the author of more than 80 publications and 60 patents. He has received a number of awards, including the 1995 Earle B. Barnes Award from the American Chemical Society for leadership in chemical research management and the 1997 Philadelphia Catalysis Society Award for excellence in catalysis.

James A. Spearot was appointed director of the Chemical and Environmental Sciences Laboratory at the General Motors (GM) Research and Development Center in August 1998. His laboratory's mission is to develop cost-effective environmental strategies and systems for General Motors' products and processes. Key research areas for the laboratory include life-cycle analysis, low-cost emissions control strategies, environmental systems for advanced material processing, fuel and lubricant systems for advanced powertrains, and innovative, efficient test environments and analytical measurements. Additionally, Dr. Spearot serves as chief scientist of GM's Powertrain Division, a position he has held since November 1998.

A native of Hartford, Connecticut, Dr. Spearot was born on April 26, 1945. He received a bachelor of science degree in chemical engineering from Syracuse University in 1967 and a master's and doctorate, also in chemical engineering, from the University of Delaware, in 1970 and 1972, respectively.

Dr. Spearot began his GM career in 1972 as an assistant senior research engineer in the Fuels and Lubricants Department. He held positions of increasing responsibility, including principal research engineer and section manager of surface and rheological studies, which led to his appointment as department head in 1992.

He is a member of several organizations: the Society of Automotive Engineers (SAE), the Society of Rheology, the American Institute of Chemical Engineers, and the American Society for Testing and Materials (ASTM). He is a former chairman of the SAE Fuels and Lubricants Division and serves on the Fluids Committee of the Engine Manufacturers Association. He also serves as chairman of the Fuels Working Group of the Partnership for a New Generation of Vehicles (PNGV). His professional honors include an ASTM Award for Excellence in 1990; the Arch T. Colwell Merit Award from the SAE in 1987; and the Award for Research on Automotive Lubricants, also from the SAE, in 1987.

Dr. John Stringer is the director of materials and chemistry support in the Science and Technology Development Division at the Electric Power Research Institute (EPRI) in Palo Alto, California. In 1988 he was appointed director of technical support in the Generation and Storage Division, and he assumed his present post in 1991. From 1982 to 1988, he was manager of the Materials Support Program and also of the Exploratory Research Program.

Before joining EPRI in 1977 as project manager in the Materials Support Program, Dr. Stringer was head of the Department of Metallurgy and Materials Science at the University of Liverpool, England. From 1963 to 1966, he worked for Battelle Memorial Institute in Columbus, Ohio, as a fellow in the Metal Science Group. From 1957 to 1963, Dr. Stringer was a member of the teaching staff in the Department of Metallurgy at the University of Liverpool.

After receiving a B.S. degree in engineering with first class honors in metallurgy from the University of Liverpool in 1955, Dr. Stringer was awarded the Ph.D. degree in 1958 and a doctor of engineering degree from the university in 1975. He is the author of two books, editor of nine others, and the author or coauthor of more than 300 papers, primarily in the areas of high-temperature oxidation and corrosion of metals and alloys, galvanomagnetic effects in alloys, and erosion and corrosion of components in fluidized-bed combustors.

Dr. Stringer is a fellow of the Institute of Fuel, a fellow of the American Association for the Advancement of Science, a fellow of the Royal Society of Arts, and a chartered engineer (U.K.). He is one of the first group of fellows of NACE International (formerly the National Association of Corrosion Engineers), elected in 1993, and a fellow of the Metallurgical Society of the American Institute of Metallurgical Engineers, elected in 1992. He is also a member of the American Society for Metals and of the Materials Research Society. In 1993 he was awarded the Ulick R. Evans Award of the Institute of Corrosion (U.K.) "for outstanding work in the field of corrosion."

David C. Thomas as manager of CO_2 Mitigation Technology, leads BP Amoco's efforts in reducing CO_2 emissions from its operations. He has held a broad range of positions in technology development, research, management, and strategy development in exploration, production, and chemicals. Dr. Thomas holds a Ph.D. in physical chemistry from the University of Oklahoma and has published more than 40 papers and 5 patents.

John A. Turner, Ph.D., is a senior electrochemist in the Center for Basic Sciences at the National Renewable Energy Laboratory. His research is primarily concerned with direct conversion (photoelectrolysis) systems for hydrogen production from water. His monolithic photovoltaic-photoelectrochemical device has the highest efficiency of any direct-conversion water-splitting device (>12%). Other work involves the study of new materials for fuel cell separators, corrosion of bipolar plates (fuel cells), electrode materials for high-energy-density lithium batteries, and fundamental processes of charge transfer at semiconductor electrodes. These research projects involve electrocatalysis, new semiconductor materials, surface modification, and the development of novel experimental techniques. He is the author or coauthor of more than 50 peer-reviewed publications in the areas of photoelectrochemistry, batteries, general electrochemistry, and analytical chemistry.

C

Origin of and Information on the Chemical Sciences Roundtable

In April 1994, the American Chemical Society (ACS) held an Interactive Presidential Colloquium entitled *Shaping the Future: The Chemical Research Environment in the Next Century*.[1] The report from this colloquium identified several objectives, including the need to ensure communication on key issues among government, industry, and university representatives. The rapidly changing environment in the United States for science and technology has created a number of stresses on the chemical enterprise. The stresses are particularly important with regard to the chemical industry, which is a major segment of U.S. industry; makes a strong, positive contribution to the U.S. balance of trade; and provides major employment opportunities for a technical work force. A neutral and credible forum for communication among all segments of the enterprise could enhance the future well-being of chemical science and technology.

After the report was issued, a formal request for such a roundtable activity was transmitted to Dr. Bruce M. Alberts, chairman of the National Research Council (NRC), by the Federal Interagency Chemistry Representatives, an informal organization of representatives from the various federal agencies that support chemical research. As part of the NRC, the Board on Chemical Sciences and Technology (BCST) can provide an intellectual focus on issues and fundamentals of science and technology across the broad fields of chemistry and chemical engineering. In the winter of 1996, Dr. Alberts asked BCST to establish the Chemical Sciences Roundtable to provide a mechanism for initiating and maintaining the dialogue envisioned in the ACS report.

The mission of the Chemical Sciences Roundtable is to provide a science-oriented, apolitical forum to enhance understanding of the critical issues in chemical science and technology affecting the government, industrial, and academic sectors. To support this mission, the Chemical Sciences Roundtable does the following:

---

[1]*Shaping the Future: The Chemical Research Environment in the Next Century*, American Chemical Society Report from the Interactive Presidential Colloquium, April 7-9, 1994, Washington, D.C.

- Identify topics of importance to the chemical science and technology community by holding periodic discussions and presentations and gathering input from the broadest possible set of constituencies involved in chemical science and technology.
- Organize workshops and symposia and publish reports on topics important to the continuing health and advancement of chemical science and technology.
- Disseminate the information and knowledge gained in the workshops and reports to the chemical science and technology community through discussions with, presentations to, and engagement of other forums and organizations.
- Bring topics deserving further, in-depth study to the attention of the NRC's Board on Chemical Sciences and Technology. The roundtable itself will not attempt to resolve the issues and problems that it identifies—it will not make recommendations nor provide any specific guidance. Rather, the goal of the roundtable is to ensure a full and meaningful discussion of the identified topics so that the participants in the workshops and the community as a whole can determine the best courses of action.

D

Acronyms and Definitions

Annex 1	Developed countries that agreed to emissions commitments in the Kyoto Protocol negotiations
AOG	Abundant oil and gas; a reference energy system
ATS	Advanced turbine systems
Carbon intensity	Carbon per unit energy
CANDU	Canada Deuterium Uranium
CBF	Coal bridge to the future—a reference energy system
DOE	Department of Energy
EIA	U.S. Energy Information Administration
EOR	Enhanced oil recovery
EPRI	Electric Power Research Institute
ESP	Electrostatic precipitation
FCC	Fluid catalytic cracker
FCCC	See UNFCCC
FGC	Fuel gas desulfurization
GDP	Gross domestic product
GMO	Genetically modified organisms
GNP	Gross national (or world) product
GRAS	Generally recognized as safe
GtC	Gigatonnes carbon (10^9 tons)
HRSG	Heat recovery steam generator
ICM	Industrial Carbon Management
IEA	International Energy Agency
IGCC	Integrated gasification combined cycle
IPCC	Intergovernmental Panel on Climate Change

APPENDIX D

IS92a,b,...	Future scenerios of IPCC based on assumptions, the more significant of which are population, population growth, rate of end-use energy intensity improvement, and elasticity of energy demand
LCI	Life cycle inventory
Mw(e)	Megawatts of electric power
NGO	Non-governmental organization
OECD	Organization for Economic Cooperation and Development
OGF	Same as AOG
oxyfuel	Fuel burned in oxygen that has been separated from air prior to burning.
PCC	Post-combustion capture
PCDC	Pre-combustion decarbonization
PET	polyethylene teraphthalate
PLA	Polylactic acid
PNGV	Partnership for a New Generation of Vehicles
ppmv	Parts per million volume
PSI	Pounds per square inch
PV	Photovoltaic
Quad	10^{15} BTU. This is approximately equal to 0.2 billion tonnes of carbon
SCFD	Standard cubic feet per day
ton	2000 lb, used by some writers. See tonne.
tonne	Metric ton, equal to 2202 lb. The standard unit for projection is tonnes of carbon. This should be differentiated from tonnes of CO_2, which is a factor of 11/3 larger than tonnes carbon
UNFCCC	United Nations Framework Convention on Climate Change. Also FCCC